WITHDRAWN
UTSA LIBRARIES

RENEWALS 458-4574

Elementary Signal Detection Theory

Elementary Signal Detection Theory

Thomas D. Wickens
University of California, Los Angeles

OXFORD
UNIVERSITY PRESS

OXFORD

UNIVERSITY PRESS

Oxford   New York

Athens   Auckland   Bangkok   Bogotá   Buenos Aires   Cape Town
Chennai   Dar es Salaam   Delhi   Florence   Hong Kong   Istanbul   Karachi
Kolkata   Kuala Lumpur   Madrid   Melbourne   Mexico City   Mumbai   Nairobi
Paris   São Paulo   Shanghai   Singapore   Taipei   Tokyo   Toronto   Warsaw

and associated companies in
Berlin   Ibadan

Copyright © 2002 by Oxford University Press, Inc.

Published by Oxford University Press, Inc.
198 Madison Avenue, New York, New York 10016

Oxford is a registered trademark of Oxford University Press

All rights reserved. No part of this publication may be reproduced,
stored in a retrieval system, or transmitted, in any form or by any means,
electronic, mechanical, photocopying, recording, or otherwise,
without the prior permission of Oxford University Press.

Library of Congress Cataloging-in-Publication Data
Wickens, Thomas D., 1942–
Elementary signal detection theory / Thomas D. Wickens.
p.   cm.
Includes bibliographical references and index.
ISBN 0-19-509250-3
1. Signal theory   2. Signal detection.
I. Title.
TK5102.5. W5387   2001
621.383'23—dc21   00-066540

9 8 7 6 5 4 3 2 1

Printed in the United States of America
on acid-free paper

Library
University of Texas
at San Antonio

Preface

Consider what happens when a person must decide whether or not an event has occurred, using information that is insufficient to completely determine the correct answer. Such situations are common—examples are given below and in Chapter 1. Without sufficient information, some errors are inevitably made. How often errors are made and their particular form depends both on the nature of the event to be detected and on how the person decides which response to make. Signal-detection theory is a widely-used method for analyzing this type of situation and of separating characteristics of the signal from those of the person who is detecting it. This book is an introduction to the theory.

The prototypical signal-detection situation is perceptual. The person, usually known as the observer, is presented with a noisy stimulus and must decide if that stimulus contains a weak signal. For example, an observer could be asked to decide whether a very faint tone had sounded in a noisy environment. Because the tone is weak and the noise loud, the observer makes errors, sometimes of omission (failing to report the tone when it had sounded) and sometimes of commission (reporting the note when it wasn't there). Although the original applications of the theory were to such perceptual decisions, it is much more widely useful. The same principles apply to decisions in many other domains—tests of memory and the theory of medical decisions are two important instances., the signals here being the memory of an event and the occurrence of a disease.

Before starting out, I want to make a few general points about the theory and its applications. Signal-detection theory is not a single monolithic method of analysis. It encompasses a range of different representations, appropriate to different tasks or sets of assumptions about detection. These different descriptions are linked by a common treatment of the decision process, and it is this general approach that constitutes signal-detection theory in the broad sense. To make particular representations of tasks, situations, observers, and so forth distinct from the overall approach, I will refer to them as *signal-detection models* and keep the term signal-detection theory for the approach as a whole. These models differ in various ways, sometimes in the observer's task, sometimes in the observer's ability to accumulate information about the stimulus, and sometimes in the way in which the observer processes that information.

The many forms of signal-detection theory serve several overlapping, but not identical, functions. Perhaps the most important of these is as a data analysis technique. For reasons that are described below, the simplest descriptions of the observer's performance in a detection task, such as the probability of a correct responses or the number of times that a signal is detected when it is presented, are imperfect summaries of what is going on. An analysis based on the signal-detection model provides better measures of such things as the facility with which the observer detects the signal.

A second, closely related, use of signal-detection theory is to order and interpret the differences among a collection of experimental conditions. To the extent that procedural differences map cleanly onto changes in the parameters of a theoretical description, these differences are more readily comprehensible. A great success of the theory is that it provides this organization for many collections of data and suggests how procedural changes manifest themselves in performance.

A third application of signal-detection theory goes further afield. The theory can be taken as a psychological model, that is, as a description of the actual process that an observer goes through when making a decision. Particularly in the form that will be discussed in Chapter 9, it gives a plausible picture of how information is combined to make a decision. This application is most open to argument. Like any description of psychological process, the signal-detection models are at best idealizations. Even when they capture the regularities of the experimental procedures, they may only describe the observer's decision process at a very general level. One should be wary (even if optimistic) of attributing too much reality to the abstract representation of the models. However, any doubts about signal-detection theory as a psychological description do not reduce its value as a way to measure performance and organize data.

The goal of this book is to introduce the reader to the most important aspects of signal-detection theory. A reader initially unfamiliar with the theory should end up able to use it to analysis data. Where I can, I have given specific recommendations. The inevitable result of this specificity is that I have ignored many of the alternative measures and competing models that have been developed over the years. A reader can find these in some of the other references cited below—a good place to start is Macmillan & Creelman, 1991. I also do not discuss some of the questions that have been raised about the validity of the signal-detection approach or the alternatives that have been proposed to it, particularly as a psychological theory.

A discussion of these alternatives requires a treatment that is at least as extensive as the one I have presented here. They are not elementary in the sense I intend this book to be, and they can only be evaluated once the basic form the theory is understood.

I have referred to the theory described in this book as elementary. There are several senses in which this description applies. As I indicated, the book presents a core description of signal-detection theory, with only limited looks at its alternatives and ramifications. I have also tried to keep it elementary the level of quantitative sophistication that is required of the reader, although not to the extent of compromising the theory. Alas, "elementary" is not synonymous with "easy." Signal-detection theory makes extensive use of probability theory, and cannot be understood without its use, except at the most superficial level. The appendix to this book contains a necessarily brief review of the important concepts as they are used here. The reader should turn to it as needed. The theory also uses the calculus in many places, often as a notational convenience, but sometimes as an analytic tool. I have used it freely in the former, less demanding, sense, but have tried to give pathways around any derivation that uses it. Mathematical symbolism that may be unfamiliar is explained in footnotes.

The contents of this book divide roughly into four parts, which differ in their generality and difficulty. A reader may choose not to study all of them on one reading. A brief description will make the structure clear. The first four chapters describe the analysis of simple detection data. Chapters 1 and 2 describe the simplest of the signal-detection models, applicable to situations like the tone study mentioned above. Stimuli are presented that may or may not contain a signal, and the observer responds by saying "yes" or "no" (or making some equivalent indication). Much real data have this form, so this version of the signal-detection model is the most widely applied. The reader of these chapters will be able to make a signal-detection analysis of such data. Chapter 3 extends this model to richer sources of data, and Chapter 4 describes the various measures of performance that are needed to accommodate this extension.

The next four chapters extend this model in various ways. Chapters 5 and 6 apply the representation developed in Chapter 3 to different experimental procedures. A reader who completes the book to this point should be equipped to understand and use most applications of signal-detection theory to detection tasks. Chapter 7 describes the changes in the model that are necessary to deal with discrimination tasks—those in which the observer must decide which of two signals is present rather than whether one is there at all. Chapter 8 introduces a new class of signal-detection models. Although it stays within the overall signal-detection framework, it treats the evidence extracted from the stimulus as essentially all-or-none instead of the graded or numerical form used in the earlier chapters.

The third part of the book recasts the decision process of signal-detection theory in much more general terms. Chapter 9 introduces the notion of a likelihood, in its technical sense, and describes how an observer can base decisions on it. The likelihood approach has several advantages over the simpler representations used in the earlier chapters. At a mathematical level, it connects signal-detection theory to the methods of statistical decision theory. By doing so, it gives a general procedure by which complex sources of evidence can be integrated into a single decision. It also eliminates some of the complexities and paradoxes that arise in simpler versions of the theory. Finally, the likelihoods give a way to interpret signal-detection theory as a psychological model that may be closer to how the observer actually makes decisions (or at least to an idealization of this operation). Personally, I find this treatment more appealing and satisfactory than that of the earlier models. The extensions provided by likelihood theory are used in Chapter 10 to describe decisions about multidimensional stimuli—those that have more than one component. These two chapters are somewhat more abstract than the previous ones, and as such they make greater mathematical demands on the reader, less at the level of the mathematical knowledge that is required than in the general level of abstraction. I would view them as essential for any reader who wishes to develop new signal-detection models or who wishes to think of the theory as a psychological construct.

The final chapter differs from the previous ten. Many forms of data analysis require statistical procedures, and, to the extent that these depend of the form of the signal-detection model, they are discussed in Chapter 11. These questions apply to all the models in this book. A reader who has decided to defer study of the middle portion of the book (especially Chapter 9 and 10) should still look over this chapter.

Although I briefly discuss computer programs for signal-detection theory in Section 3.6, I do not describe any of the existing programs in detail or give examples of their use. In part I have omitted them because I wished to focus on the theory itself. More importantly, computer resources are a swiftly moving target, and discussions of them are all to often obsolete by the time they are published. Fortunately it is no longer necessary to confound these ephemeral matters with the underlying theory. I will provide some useful computer programs, particularly those that I have developed for this book, on my web site, at `www.bol.ucla.edu/~twickens`.

This book originated in a series of informal talks that I gave to a group of graduate students in the Cognitive program at UCLA. It has taken a leisurely path from this start, moving from lecture notes to handouts in a variety of forms, and through a series of partial drafts. I have used it in classes in all these forms, as have several of my colleagues. It has been shaped, and substantially improved, by the feedback I have received. The number of students, colleagues, and friends (not mutually exclusive categories) who have given me suggestions, spotted errors, and given encouragement is so large that I cannot attempt to thank them by name here. I hope that each of you can recognize and find pleasure in your contribution. In spite of all the assistance I have received, I fear that various infelicities remain—I wish I could believe no errors, but the theory itself tells me that that is improbable. These, and the places where I have indulged my own biases in the face of advice, are my own responsibility. I would welcome any thoughts, corrections, or suggestions from readers.

Thomas D. Wickens
twickens@psych.ucla.edu
San Francisco, March 2001

Contents

Elementary Signal Detection Theory

Chapter 1

The signal-detection model

Signal-detection theory provides a general framework to describe and study decisions that are made in uncertain or ambiguous situations. It is most widely applied in psychophysics—the domain of study that investigates the relationship between a physical stimulus and its subjective or psychological effect—but the theory has implications about how any type of decision under uncertainty is made. It is among the most successful of the quantitative investigations of human performance, with both theoretical and practical applications. This chapter introduces the basic concepts of signal-detection theory, in the context of simple yes or no decisions. These ideas are expanded and amplified in later chapters.

1.1 Some examples

A good place to begin the study of signal-detection theory is with several examples of detection situations. Consider a very common situation: an individual must decide whether or not some condition is present. Such decisions are easy to make when the alternatives are obvious and the evidence is clear. However, there are many tasks that are not so simple. Often, the alternatives are distinct, but the evidence on which the decision is to be based is ambiguous. Here are three examples:

- A doctor—physician, psychologist, or whoever—is examining a patient and trying to make a diagnosis. The patient shows a set of symptoms, and the doctor tries to decide whether a particular disorder is present or not. To complicate the problem, the symptoms

3

are ambiguous, some of them pointing in the direction of the disorder, others pointing away from it; moreover, the patient is a bit confused and does not describe the symptoms clearly or consistently. The correct diagnosis is not obvious.

- A seismologist is trying to decide whether to predict that a large earthquake will occur during the next month. As it was for the doctor, the evidence is complex. Many bits of data can be brought to bear on the decision, but they are ambiguous and sometimes contradictory: records of earthquake activity in the distant past (some of which are not very accurate), seismographic records from recent times (whose form is clear, but whose interpretation is not), and so forth. The alternatives are obvious: either there will be a quake during the month or there will not. How to make the prediction is unclear.

- A witness to a crime is asked to identify a suspect. Was this person present at the time of the crime or not? The witness tries to remember the event, but the memory is unclear. It was dark, many things happened at once, and they were stressful, confusing, and poorly seen. Moreover, the crime occurred some time ago, and the witness has been interviewed repeatedly and has talked to others about the event. Again the alternatives are clear: either the person was present or not, but the information that the witness can bring to bear on the decision is ambiguous.

These three examples have an important characteristic in common. Although the basic decision is between simple alternatives—in each case, the response[1] is either YES or NO—the information on which that decision is based is incomplete, ambiguous, and frequently contradictory. These limitations make the decision difficult. No matter how assiduous the decision maker, errors will occasionally occur.

A full understanding of the decision process in situations such as these three is difficult, if not impossible. The decisions depend on many particulars: what the decider knows, his or her expectations and beliefs, how later information affects the interpretation of the original observations, and the like. An understanding of much domain-specific knowledge is needed—of medical diagnosis or seismology, for example. This base of knowledge must be coupled with a theory of information processing and problem solving. Fortunately, much can be said about the decision process without going into these details. Some characteristics of a yes-no decision transcend the particular situation in which it is made. The theory discussed here treats these common elements. This theory, known generally as *signal-detection theory*, is one of the greatest successes of mathematical psychology.

---

[1]Throughout this book, SMALL CAPITALS are used to denote overt responses.

Although signal-detection theory has implications for the type of complex decisions described above, it was developed for a much simpler situation: the detection of a weak signal occurring in a noisy environment. It is much easier to describe the theory in the simpler context. Consider two experiments:

- *Classical signal detection.* A person, referred to as the *observer*, is listening through earphones to white noise.[2] The observer also is watching a small light. At regular intervals the light goes on for one second. During this one-second interval one of two things happens: either the white noise continues as before, or a faint 523 Hz tone (middle C on the musical scale) is added to the noise. Whether the tone is added or not is decided randomly, so the observer has no way to tell in advance which event will occur. At the end of the one-second interval the stimulus returns to normal (if it had changed). The observer decides whether the tone was presented during the lighted interval and reports the choice by pressing one of two buttons, the left button if no tone was heard, the right button if a tone was heard. Because the tone is weak and the background noise loud, it is easy for the observer to make an error in detecting the tone.

- *Recognition memory.* The subject in this recognition-memory experiment is seated in a darkened room and views a series of slides projected on a screen. The slides are photographs of various outdoor scenes—trees, waterfalls, buildings, and so forth. Each slide is shown for three seconds, providing time to get a look at it but not to study it carefully. After 200 slides have been seen, the subject works on an unrelated task for half an hour, then returns to the room and is given a memory test. A new series of slides is shown. Half of these slides are old pictures, drawn from the 200 that were seen before. The other half are new pictures, generally similar to the old slides, but of novel scenes. For each of these slides the subject is asked to decide whether it had been seen in the first series or not.

If one overlooks the particular differences—perception or memory, audition or vision, meaningless tones or meaningful pictures—these two tasks are much alike. In each, a well-defined yes-no decision is made (tone present or absent in one case, picture old or new in the other), in each the signal that the observer is trying to detect is weak (the faint tone or the memory of a briefly seen picture), and in each the task is complicated by interfering information (the white noise or the competing memories of similar pictures

---

[2]This type of sound is a random mixture of a wide range of frequencies. It is called "white" by analogy with white light, which is a mixture of all colors. White noise sounds much like the hissing noise one hears from a radio that is not tuned to any station.

seen in the past). The common structure lets the two tasks be analyzed in similar ways.

1.2   Hits and false alarms

To start out, some terminology is needed. The observer in a detection experiment experiences two types of trials. On some trials only the random background environment is present, either the white noise or the random familiarity of a new picture. Because they contain no systematic component, these trials are called *noise trials*. On other trials, some sort of signal is added to the noise. In the examples, the signal is either the tone or the added familiarity associated with having recently seen a picture. These trials are called *signal plus noise trials* or, more simply, *signal trials*. When speaking of the recognition memory experiment, these possibilities correspond to *new items* and *old items*, respectively. In the discussion below, the signal-and-noise formulation will be used. Whichever terminology is used, the observer makes either the response YES or the response NO.

At first glance, it seems easy to score experiments of this type. Just measure how well the observer does by finding the proportion of times that the signal is detected. The event in question, saying YES to a signal, is known as a *hit*. The proportion of hits is calculated by dividing the frequency of hits by the frequency of signal trials to give the *hit rate*:

$$h = \frac{\text{Number of YES responses to signals}}{\text{Number of signal trials}}.$$

Good observers or observers presented with easy signals have high hit rates and poor observers or ones presented with difficult signals have low hit rates.

Unfortunately, the hit rate is incomplete as a summary of these experiments and is not a good way to indicate how well the signal is detected. It depends, in part, on aspects of the task other than the detectability. Consider one observer in two sessions of the tone detection experiment. During each session the observer is presented with 100 trials on which a pure noise stimulus is present and 100 trials on which the tone is added to the noise in an appropriate random sequence. The same signal is used in both sessions, so that the ability of the observer to detect it should be the same (at least as long as there are no effects of practice). The sessions differ in the instructions given to the observer. In the first session, the observer is told that it is very important to catch all the signals. To give incentive, the observer is paid 10 cents for every correctly detected signal. When the data for the session are analyzed, 82 hits have been made on 100 signal trials, so the proportion of hits is $h = 0.82$. The second session has different instructions. Now the observer is told that it is very important not to report a signal when there isn't one, a type of error known as a *false alarm*. To emphasize these instructions, the bonus for hits is removed, and the observer receives 10 cents for every correctly identified noise trial. There are far fewer hits in this session: only 55 of the 100 signal trials receive YES responses, giving $h = 0.55$. These results suggest that the proportion of hits is unsatisfactory, or at least incomplete, as a measure of the properties of the signal. Were it really a good measure, its value would be the same in both sessions—remember that the signal itself does not change. However, the proportion of hits drops from 0.82 to 0.55.

It is easy to see informally what happened here. The signals were weak and easily confused with the noise. On some trials this confusion made the observer unsure what to answer. In this state of uncertainty, the behavior of the observer in the two sessions is likely to differ. In the first session, to maximize the number of hits, the observer does best to say YES when unsure. After all, if that choice is correct, then the reward is received, while if it is wrong, nothing bad happens. In contrast, during the second session it is best to say NO when uncertain. With this response, if there really were no signal, then the reward would be received, while if there were a signal, nothing bad would happen. The detectability of the signal has not changed, but manipulating the payoff has made the observer change strategy, altering the proportion of hits. A better measure of detectability should not be affected by these strategic matters.

As this example shows, the problem with an analysis based only on the hit rate is that it neglects what happens on trials where the noise stimulus is presented. A complete picture requires attention to both possibilities. There are two types of trial, either noise or signal, and there are two responses, either NO or YES. Because each type of response can occur with each type of stimulus, there are four possible outcomes. These four possibilities are identified by name:

		Response	
		NO	YES
Trial Type	Noise	Correct Rejection	False Alarm
	Signal	Miss	Hit

Two of these possibilities, hits and correct rejections, are correct; the other two, false alarms and misses, are errors.

An idea of what is lost by looking only at the hits is seen by examining the complete data for the experiment described above. Suppose that in the two sessions the frequencies of the four events were

	First Session				Second session	
	NO	YES			NO	YES
Noise	54	46	and	Noise	81	19
Signal	18	82		Signal	45	55

The decrease in hits between the first and the second session here is matched by a decrease in false alarms. Because the observer was rewarded for hits in the first session and for correct rejections (i.e., for avoiding false alarms) in the second session, this behavior is quite appropriate.

Parenthetically, note that this observer does not behave in a way that maximizes the earnings. There was no reward for correct rejections or penalty for false alarms in the first session, so the observer could earn the maximum possible amount money by saying YES on every trial. Likewise, the highest paying strategy in the second session is always to say NO. Had these strategies been used, the two tables of data would have appeared

	First Session				Second session	
	NO	YES			NO	YES
Noise	0	100	and	Noise	100	0
Signal	0	100		Signal	100	0

However, most observers are loath to adopt such extreme strategies, and data of the type shown earlier, which are biased in the correct direction but less exaggerated, are more common.

Although there are four types of outcome here, one does not need four numbers to summarize the observer's behavior. In these detection experiments, the frequencies of the two types of trials—that is, of noise trials and signal trials—are determined by the experimenter. The observer's behavior governs the proportion of YES and NO responses on trials of a single type, not how many trials there are. When describing the observer's behavior, one does not want to summarize the data in a way that partially reflects the experimenter's behavior. So the results are usually converted to conditional proportions in each row. Conventionally in signal-detection theory, the two probabilities used are the *hit rate*,

$$h = \frac{\text{Number of hits}}{\text{Number of signal trials}},$$

and the *false-alarm rate*,

$$f = \frac{\text{Number of false alarms}}{\text{Number of noise trials}}.$$

One can also calculate a miss rate and a correct rejection rate, but these quantities are redundant with the hit rate and the false-alarm rate:

$$\text{miss rate} = 1 - h,$$
$$\text{correct rejection rate} = 1 - f.$$

These two proportions add no new information to that provided by the hit rate and false-alarm rate, so it has become conventional to report only h and f. Using these two statistics, the data from the two sessions are summarized as

	h	f
First Session	0.82	0.46
Second Session	0.55	0.19

The way that the hit rate and the false-alarm rate increase or decrease together between the two sessions is quite apparent. This table contains the minimum information that must be reported to understand what is happening in a pair of two-alternative detection experiments.

The ideas expressed here have been developed in many domains, and with them other terminology. A more neutral nomenclature refers to hits and false alarms as *true positives* and *false positives*, and to correction rejections and misses as *true negatives* and *false negatives*. In the epidemiological literature, the *sensitivity* of a test is its hit rate, and the *specificity* is its correct-rejection rate.

1.3 The statistical decision representation

Although reporting both the hit rate and the false-alarm rate is much better than presenting one of these alone, even together they are not completely satisfactory. Together the two proportions give an idea of what the observer is doing, but neither number unambiguously measures the observer's ability to detect the signal. Neither the hit rate nor the false-alarm rate tells the whole story. A single number that represents the observer's sensitivity to the signal is better. The theories discussed in this book provide this measure.

To develop a measure of sensitivity, it is necessary to go beyond a simple description of the data. A measure that describes the detectability of a signal must be based on some idea of how the detection process works as a whole. A conceptual picture of the detection process, known broadly as *signal-detection theory*, has been developed to provide this structure. There are several ways that this broad picture can be made more specific and given a rigorous mathematical form. Each of these signal-detection *models*, as they will be called, leads to a different way to measure detectability. The simplest of these models describes an observer's performance by a pair of numerical quantities, one that measures the detectability of the signal and another that measures the observer's preference for YES or NO responses. Using this model, the data quantities h and f are translated into estimates of more meaningful theoretical quantities. This section describes one model for the two-alternative case; it is developed, extended, and modified in later chapters.

The basic model is drawn from statistical decision theory and is similar to the ideas that are used in statistical testing to make a decision between two hypotheses. This theory, as it is applied in signal-detection theory, is founded on three assumptions:

1. The evidence about the signal that the observer extracts from the stimulus can be represented by a single number.
2. The evidence that is extracted is subject to random variation.
3. The choice of response is made by applying a simple decision criterion to the magnitude of the evidence.

The next few paragraphs describe these assumptions in more detail.

First, consider the underlying dimension. The idea here is that the internal response of the observer to a stimulus, insofar as the detection decision is concerned, can be represented by a point on a single underlying continuous dimension. For example, when the signal is a pure tone of known frequency masked in white noise, this dimension might be the output of whatever neuron (or set of neurons) in the auditory system is maximally responsive to the appropriate signal frequency. For the recognition memory experiment, this dimension is some feeling of "familiarity" with the test word, perhaps as contrasted with a feeling for how familiar a new word would be. A more abstract definition is provided by the concept of a *likelihood ratio*, drawn from statistical theory, which is developed in Chapter 9.

In both examples a much more thorough analysis of specific properties of the detection system is possible. One could look (as many researchers have done) at the detailed neurology of tone detection or the various visual and semantic dimensions that influence recognition memory. However, this detail is unnecessary when seeking a detection measure. One of the powers of signal-detection theory is its ability to give useful quantitative answers without requiring one to delve into the specifics. Under the signal-detection model, the response depends on a single value, and all information obtained by the observer is summarized in this number. Thus, the theory can apply both to the complex medical or seismological examples at the start of this chapter and to the simpler perceptual or mnemonic studies.

Another important point drawn from statistical decision theory is embodied in the second assumption. Trials differ in their effect, even when the nominal stimulus is the same. Sometimes this variability has a physical interpretation. For the tone-detection experiment, the white noise is a random mixture of energy at all different frequencies; thus on a given noise trial the output of a physical detector tuned to 523 Hz (if that is the location of the signal) will be larger or smaller, depending on the accidental composition of the noise. On signal trials the added tone gives an increment to the output of the detector, but the variability attributable to the noise is still there. In other situations the variability, although no less present, lacks this clear physical interpretation. In the recognition memory experiment a given picture evokes a greater or lesser feeling of familiarity, depending on how often similar pictures have been seen in the past and under what circumstances, and, for old pictures, how attentive the subject was during the original presentation. Even if the physical basis for this variability is less apparent, its effect is no less true. Sometimes a new picture seems very familiar; sometimes an old one seems less so.

The variation of the stimulus effect is well illustrated by a diagram that shows the distributions of evidence under the two alternatives (Figure 1.1). The horizontal axis is the single dimension on which the internal response to a stimulus is measured, and the vertical height of the line indicates how likely that value of the evidence is to occur. The top curve, or distribution of evidence, describes the internal response when a noise-only trial is given, and the bottom curve describes the response to the signal-plus-noise stimulus. The two distributions are not identical, indicating that the observations are, to some degree at least, sensitive to the signal— if they were identical, then there would be no way to tell the two events apart. The presence of a signal changes both the center and the spread of the distribution, both its mean and its standard deviation in statistical terms. Most important, the signal distribution is shifted to the right relative to the noise distribution. On average, larger values are observed when a signal is present than when it is absent. However, because the distributions overlap, some noise trials produce a larger observation than some signal trials. For example, both the noise distribution and the signal distribution have positive values at the point marked x in Figure 1.1, implying that there is some chance that evidence of that strength is observed, whichever type of stimulus is presented. This ambiguity means that no response rule based on this dimension can produce completely error-free performance.

The two distributions drawn in Figure 1.1 differ in other ways than in the mean—here the distribution when the signal is present is more spread out then when it is absent. This difference is not a necessary part of the signal-detection model. In some versions of the theory, such as that discussed

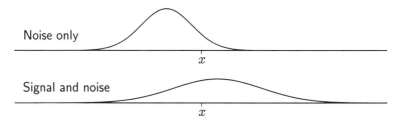

Figure 1.1: *Probability distributions on the evidence dimension for a noise stimulus (upper distribution) and for a signal-plus-noise stimulus (lower distribution). The value x represents a particular observation, which could have come from either distribution.*

in Chapter 2, the two distributions differ only in their center. In other version, such as those discussed later in this book, variability differences are also possible. Methods for testing which possibility is most appropriate are described in Chapter 3.

The third assumption of signal-detection theory links the abstract dimension of the internal response to the observer's overt dichotomous response. The linking is very simple: on any trial, the observer says YES when the amount of evidence for the signal is larger than some value known as the *criterion*, and NO when it is smaller than this value. The observer's *decision rule* to determine the response is

$$\begin{cases} \text{If evidence} > \text{criterion, then respond YES,} \\ \text{If evidence} < \text{criterion, then respond NO.} \end{cases}$$

When the evidence is assumed to be a continuous variable, there is no need to worry about whether the evidence exactly equals the criterion, as the probability of that event is negligible, and it can be assigned to either the YES or NO category without changing the properties of the rule. This decision rule is pictured in Figure 1.2. The point marked λ (the lowercase Greek letter lambda) on the abscissa is the criterion that divides the response types. For any value of X above λ, a response of YES is made; for any value of X below λ, a response of NO is made.

For compactness, Figure 1.2 (unlike Figure 1.1) has been drawn with both distributions on the same axis. This condensed diagram is the conventional way to present the signal-detection model. However, one should remember that each actual event is drawn from one or another of these distributions. On any trial, only one of them applies.

Some mathematical notation is needed to take the analysis further. Denote values on the abscissa of Figure 1.2 by the letter x. Under the signal-

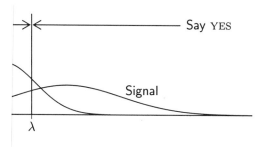

Say YES

Signal

λ

...oise distributions of Figure 1.1 shown on a single
at the value λ.

...t of evidence observed on a single trial is not a
...a variable (see Appendix Section A.2). Denote
...$_n$ for noise trials and X_s for signal trials. Let
...nsity functions of these two random variables—
...ed in Figures 1.1 and 1.2—and let $F_n(x)$ and
...distribution functions. Using this notation, the
probabilities of a hit or a false alarm, given the appropriate stimulus, are
found by calculating the area under the density functions above the value
λ. In mathematical nomenclature, this area is the *integral* of the function.[3]
The false-alarm rate P_F is the probability that an observation from X_n
exceeds λ, which can be written in various ways as

$$P_F = P(\text{YES}|\text{noise})$$
$$= P(X > \lambda|\text{noise})$$
$$= P(X_n > \lambda)$$
$$= \int_{\lambda}^{\infty} f_n(x)\, dx$$
$$= 1 - F_n(\lambda). \tag{1.1}$$

The shaded area in the upper distribution in Figure 1.3 corresponds to this
probability. The hit rate P_H is defined similarly, using the distribution of

---

[3]This book uses the integral symbol from calculus to denote area. The area under
the function $f(x)$ between $x = a$ and $x = b$ is written $\int_{a}^{b} f(x)\, dx$. The important parts of
the expression are the function to be integrated $f(x)$ and the limits a and b; the rest is,
in effect, conventional notation. When $a = -\infty$, the integral includes all the area below
b, and when $b = \infty$, it includes all the area above a.

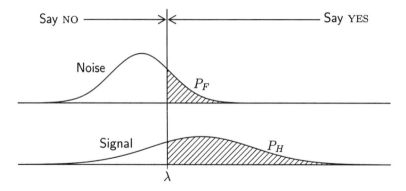

Figure 1.3: *Probabilities of a false alarm and of a hit shown as shaded areas on the distributions of X_n and X_s from Figures 1.1 and 1.2.*

evidence associated with the signal (Figure 1.3, bottom):

$$P_H = P(X_s > \lambda) = \int_\lambda^\infty f_s(x)\,dx = 1 - F_s(\lambda). \tag{1.2}$$

The probabilities P_F and P_H are the theoretical counterparts of the proportions f and h calculated from data.

The definition of the hit rate and the false-alarm rate in terms of the distributions of X_s and X_n allows predictions of these values to be made once the forms of the random variables are specified. Before turning to these calculations in the next chapter, note that two aspects of the signal-detection model control the particular values of P_H and P_F:

- The overlap in the distributions. If $f_n(x)$ and $f_s(x)$ do not overlap much, then the hit rate can be high and the false-alarm rate can be low at the same time. If the distributions are nearly the same, then the hit rate and the false-alarm rate are similar.
- The placement of the criterion. If λ lies toward the left of the distributions, then both the hit rate and the false-alarm rate are large. If λ is lies toward the right, then both rates are small.

The importance of the signal-detection model is that it allows these two aspects of the detection situation to be measured separately. The detectability of the signal is expressed by the position of the distributions and their degree of overlap, while the observer's strategy is expressed by the criterion. The shape and position of the distributions in a particular detection task may be largely determined by the way that the stimulus is generated and the physiology of the detection process. The observer has

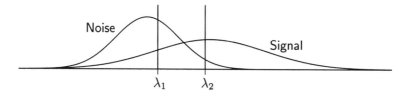

Figure 1.4: *Signal and noise distributions with two criteria λ_1 and λ_2 representing the differences induced by instructions.*

little or no control over these aspects. However, the observer can vary his or her propensity to say YES by changing the position of the criterion. Such changes can explain the differences in hit rate and false-alarm rate between the two sessions in the experiment described above. The criterion λ_1 shown in Figure 1.4 yields large hit rates and false-alarm rates, such as those in session 1, and the criterion λ_2 yields smaller rates such as those in session 2. Of course, one can only tell whether criterion differences are enough to explain the differences between the conditions by fitting the data quantitatively, which is the topic of the next two chapters.

Reference notes

Signal-detection theory, as discussed here, was originally developed in work by Birdsall, Swets, and Tanner, although its roots go back at least to Gustav Fechner's *Elemente der Psychophysik* in 1860 (see Link, 1992). Many of the early articles relating to this development are collected in Swets (1964). The primary source for signal-detection theory is the book by Green and Swets (1966), which remains an essential (but sometimes difficult) reference. There are many secondary treatments of signal-detection theory directed toward different research domains. Among those oriented to psychophysics, the introduction by McNicol (1972) and the more mathematical treatment by Falmagne (1985) have influenced the present writing. Macmillan and Creelman (1991) provide a useful and more detailed treatment. John Swets has written a number of useful articles on the procedures, many of which are collected in Swets (1996). Briefer discussions of signal-detection theory appear in many texts on perception, cognitive science, and so forth. Articles that apply signal-detection theory are too numerous to cite.

Exercises

1.1. Describe two situations, other than the examples in Section 1.1, in which a decision between two clear alternatives must be made based on incomplete or ambiguous information. Identify the types of errors that could be made.

1.2. For each of the following pairs of hit and false-alarm rates, choose the pair of distributions that best describes it and locate a criterion that gives approximately those values. Remember that, by definition, the complete area under the distributions equals 1.0.

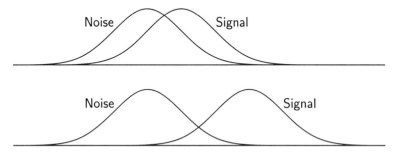

a. $P_F = 0.50$, $P_H = 0.84$.
b. $P_F = 0.20$, $P_H = 0.56$.
c. $P_F = 0.07$, $P_H = 0.93$.

1.3. Two screening tests are available that predict the appearance of a set of psychological symptoms. In a study of the first test, it is given to a group of 200 people. On a follow-up one year later, the test is found to have identified the 25 of the 30 people who developed the symptoms during the year. A comparable study using the second test looked at a different set of 150 people and identified 15 persons out of 28 who developed the symptoms. The rate of identification is greater for the first test than for the second ($25/30 = 83\%$ and $15/28 = 54\%$, respectively). Explain why these data do not, by themselves, tell which test is best at identifying the target people. What information is missing?

Chapter 2

The equal-variance Gaussian model

To make predictions from the signal-detection model described in the last chapter, the form of the distributions of X_s and X_n must be specified. One of the simplest and most natural choices is the conventional "bell-shaped" *normal distribution* of statistics, more commonly known in signal-detection applications as the *Gaussian distribution*. The next three chapters describe various uses of the Gaussian model, and it is central to much of the remainder of this book. The properties of this distribution, including methods used to find probabilities from it, are described in the appendix, starting on page 237. This material should be reviewed, either now or as required while reading this chapter.

The simplest Gaussian model is one in which the distributions for both signal and noise stimuli have the same shape, with the signal distribution being shifted to the right of the noise distribution, but otherwise identical to it. This representation is particularly useful as a description of a single detection condition and is the basis of the most commonly used detection statistics. It is the topic of this chapter.

2.1   The Gaussian detection model

In its most general form, the Gaussian signal-detection model assigns arbitrary Gaussian distributions with (potentially) different means and variances for the random variables representing the two stimuli:

$$X_n \sim \mathcal{N}(\mu_n, \sigma_n^2) \qquad \text{and} \qquad X_s \sim \mathcal{N}(\mu_s, \sigma_s^2).$$

17

The result is a model that depends on the values of five real-number parameters instead of on two arbitrary distributions. The two parameters μ_n and μ_s locate the centers of the distributions, the two parameters σ_n^2 and σ_s^2 give their variances, and the parameter λ specifies the response criterion. Once these five values are assigned, predictions of P_H and P_F can be made.

The signal-detection model does not use all five of these parameters. The values of the random variables X_n and X_s are unobservable. Only the shape and overlap of the distributions are important; the picture is unspecified as to its origin and scale. There is no way to determine the absolute size of X_n and X_s, only how they compare to each other. To illustrate this indeterminateness, consider Figure 1.2 on page 13. A zero point can be placed anywhere on the abscissa and scale values (1, 2, etc.) marked off at any spacing along the axis without changing the essential picture. The overlap of the distributions and the relative placement of the criterion are the same for any scaling. Thus, a set of detection data does not give enough information to assign unique values to the five parameters.[1]

To begin to remove the ambiguity, the values of any two of the five parameters (except the two variances) can be chosen arbitrarily. Fixing these parameters does not make the model less general or reduce its range of predictions; it only helps to give the other parameters unique values. One conventional way to fix two parameters is to assign a standard Gaussian distribution to the noise distribution, by setting μ_n to zero and σ_n^2 to one, giving the pair of distributions

$$X_n \sim \mathcal{N}(0,1) \quad \text{and} \quad X_s \sim \mathcal{N}(\mu_s, \sigma_s^2). \tag{2.1}$$

With this restriction, the model depends on three parameters, μ_s, σ_s^2, and λ.

Example 2.1: Suppose that $X_n \sim \mathcal{N}(0,1)$, that $X_s \sim \mathcal{N}(1.5, 4.0)$, and that $\lambda = 1.2$. What are P_H and P_F?

Solution: The noise distribution has standard form, so the false-alarm rate is obtained by looking up the position of the criterion in a Gaussian distribution table. Using Equation 1.1 and the table of $\Phi(z)$ on page 249,

$$P_F = P(X_n > \lambda)$$
$$= 1 - P(X_n \le \lambda)$$
$$= 1 - \Phi(\lambda)$$

---

[1] In technical terms, the fully parameterized model is not *identifiable*. Many different combinations of μ_n, σ_n^2, μ_s, σ_s^2, and λ lead to the same values of P_H and P_F; thus, there is no way to identify which combination of parameter values created a particular pair of probabilities.

$$= 1 - \Phi(1.2)$$
$$= 1 - 0.885 = 0.115.$$

The signal distribution does not have standard form, so to calculate the hit rate (i.e., to evaluate Equation 1.2), both the mean and the standard distribution must be rescaled using Equation A.46:

$$P_H = P(X_s > \lambda)$$
$$= 1 - P(X_s \leq \lambda)$$
$$= 1 - \Phi\left(\frac{\lambda - \mu_s}{\sigma_s}\right)$$
$$= 1 - \Phi\left(\frac{1.2 - 1.5}{2.0}\right)$$
$$= 1 - \Phi(-0.15)$$
$$= 1 - 0.440 = 0.560.$$

Setting the noise distribution to standard form is but one way to constrain the signal-detection model and give the parameters unique values. Another possibility is to put the origin exactly between the two distributions and equate their variances, so that

$$X_n \sim \mathcal{N}(-\mu, \sigma^2) \quad \text{and} \quad X_s \sim \mathcal{N}(\mu, \sigma^2).$$

This representation is more natural in situations where neither stimulus event can be singled out to provide a baseline, as was done with the noise distribution in simple detection. This form of model will be important in the descriptions of forced-choice and discrimination studies in Chapters 6 and 7, and it is a natural consequence of the likelihood-ratio approach described in Chapter 9.

One might ask at this point why Gaussian distributions have been used for X_n and X_s, when other distributions could work as well. One reason is familiarity: the Gaussian is the most commonly used distribution in signal-detection theory models, and in some treatments is presented as if it were the only signal-detection model. The choice of a Gaussian distribution can be justified on either empirical or theoretical grounds. Empirically, it is often supported by data, using the methods that are described in Section 3.4. Theoretically, one can argue that when the evidence being incorporated in the decision derives from many similar sources, the central limit theorem (described on page 241) implies that their combination has approximately a Gaussian form. Such a situation might obtain, for example, if X is formed by summing the outputs of a set of detectors, each of which responds to some aspect of the signal, or by pooling many nearly ambiguous pieces of evidence.

Notwithstanding these arguments, there are some situations in which distributions that are not Gaussian are necessary. The most important of these cases occur when a non-Gaussian form for the distribution arises out of a theoretical model for the detection task. Distributions very much unlike the Gaussian form occur in Chapter 8 (see Figure 8.2 on page 136) and Section 9.4 (see Figure 9.5 on page 168). In such cases, of course, it is essential to use the correct distribution.

Beyond these specific situations, one sometimes finds treatments of signal-detection theory in which X_n and X_s have what is known as a *logistic distribution* (page 241). This distribution, like the Gaussian, is unimodal, symmetrical, and bell-shaped. Its shape is so similar to that of the Gaussian distribution that it is indistinguishable from it in practical applications. The logistic distribution arises naturally out of certain axiomatic treatments of choice behavior, and when working with these theories, it is the natural form for X_n and X_s. Another reason to use it is mathematical tractability: the cumulative logistic distribution function can be written as an algebraic expression, while the cumulative Gaussian distribution function $\Phi(z)$ must be approximated or taken from tables. This difference has no practical consequence for someone applying the signal-detection model, particularly when the calculations are done by a computer. In view of its wider use, the Gaussian model is emphasized in this book.

2.2   The equal-variance model

The most common application of signal-detection theory is to a single detection condition. The data obtained from a two-alternative detection experiment consist of one hit rate h and one false-alarm rate f. By themselves, these two numbers cannot determine unique values for three parameters μ_s, σ_s^2, and λ. One more constraint is needed to fit the signal-detection model.

The variance σ_s^2 is the parameter with the least obvious interpretation, so the most natural way to restrict the model is to constrain its value. The conventional assumption is that the variance of the signal distribution is the same as the variance of the noise distribution. Setting $\sigma_s^2 = \sigma_n^2 = 1$ gives the *equal-variance Gaussian model*. The standard notation for this model denotes the mean of the signal distribution by the symbol d' rather than μ_s. Thus, the two random variables have the distributions

$$X_n \sim \mathcal{N}(0,1) \qquad \text{and} \qquad X_s \sim \mathcal{N}(d',1).$$

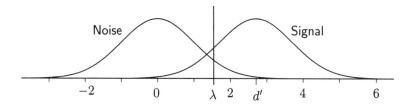

Figure 2.1: *The distribution of noise and signal for the equal-variance Gaussian signal-detection model.*

These distributions are shown in Figure 2.1. Using Equations 1.1 and 1.2 with the Gaussian distribution function, the probabilities of a false alarm and a hit are

$$P_F = 1 - \Phi(\lambda) \qquad \text{and} \qquad P_H = 1 - \Phi(\lambda - d'). \qquad (2.2)$$

This model depends on two parameters, d' and λ, which can be estimated from observations of f and h.

The two parameters of the equal-variance Gaussian detection model have a simple and very important interpretation. The parameter d' describes the relationship of the noise and signal distributions to each other. When d' is near zero, the distributions are nearly identical; when it is large, they are widely separated. Thus, d' measures how readily the signal can be detected. The parameter λ describes the position of the decision criterion adopted by the observer. It is influenced by any propensity to say YES or NO, although it is not as satisfactory a measure of response bias as the quantities that will be described in Section 2.4.

Before going on to describe how f and h are converted to estimates of d' and λ, it is worth pausing to reconsider the restrictions that have been imposed on the general signal-detection model. There are three of these, each with a somewhat different status. The first is the use of the Gaussian distribution for X_n and X_s. As discussed at the end of the last section, this distribution is a reasonable choice for most sets of detection data. In practice, one rarely worries much about this assumption.

The second restriction fixes the parameters of the noise distribution, giving it standard form, with $\mu_n = 0$ and $\sigma_n^2 = 1$. This assumption has a different status from the others. It has no real content, in that it does not restrict the range of predictions that the model can make. The unobservable variables X_n and X_s are hypothetical, and they can be given any center and spread as long as they maintain the same relationship to each other. Thus, this restriction is made purely for numerical convenience. Either this assumption or an equivalent one is necessary if well-defined parameter estimates are to be made, but it cannot be tested by any data that could be collected.

The third assumption assigns equal variance to the two distributions, putting $\sigma_s^2 = \sigma_n^2$. This assumption is the most vulnerable of the three and the least likely to be correct. It can be tested, although not with the results of a single two-alternative detection experiment—the two data values from this experiment can be fitted perfectly by an appropriate choice of d' and λ. Unlike the standardization of X_n, the equal-variance assumption has implications for data in more elaborate experiments, a topic that will be discussed in the next chapters.

2.3   Estimating d' and λ

In order to use the detection parameters d' and λ to describe an observer's performance, their values must be calculated from the observed quantities h and f, a process known as *estimation*. Estimates of d' and λ should be chosen that match the model's predictions to the observed data. The derived values are known as *estimates*, not parameters, and are denoted here by placing a circumflex or "hat" on the symbol that denotes the theoretical parameter. For the equal-variance Gaussian model, the parameter estimates are $\widehat{d'}$ and $\widehat{\lambda}$. The hat serves to distinguish the estimates from their theoretical counterparts.

To find $\widehat{d'}$ and $\widehat{\lambda}$, one uses either a table of the Gaussian distribution or a program that calculates areas under this distribution. Essentially, these tables translate the observed rates of false alarms and hits to areas in a standard Gaussian distribution. The areas are then converted to values of $\widehat{d'}$ and $\widehat{\lambda}$. The calculation is easiest to do in three steps. Consider the data from the first session of the example experiment described in Section 1.2. The data from this experiment were

	NO	YES	
Noise	54	46	100
Signal	18	82	100

From these frequencies, the hit rate and the false-alarm rate are found:

$$h = \frac{82}{100} = 0.82 \qquad \text{and} \qquad f = \frac{46}{100} = 0.46.$$

From here, the calculation proceeds as follows.

1. The false-alarm rate and the noise distribution are used to estimate the criterion λ. It is always helpful to illustrate the calculations with

a picture. The noise distribution, with the criterion marked and the false-alarm rate shaded, is

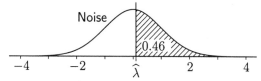

This picture shows that $\widehat{\lambda}$ is the point on a standard normal distribution above which 46% of the probability falls, or, equivalently, below which 54% of the probability falls. The latter cumulated probability is the value tabulated in most normal-distribution tables, including the one on page 250. Looking up $Z(0.540)$ in this table shows that $\widehat{\lambda} = 0.100$.

2. The hit rate is used to find the distance between d' and the criterion. A hit rate of 0.82 indicates that 82% of a Gaussian distribution centered at $\widehat{d'}$ falls above $\widehat{\lambda}$:

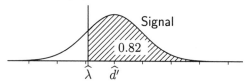

This distance is the difference between λ and d'. From the same table of $Z(p)$, 82% of the standard Gaussian distribution falls below $z = 0.915$. Converting this value to an upper-tail area, as the picture shows, $\widehat{\lambda}$ must be 0.915 units below $\widehat{d'}$:

$$\widehat{\lambda} - \widehat{d'} = -0.92.$$

3. The final step combines these two results to solve for $\widehat{d'}$. Graphically, the two diagrams are superimposed, to make the two values of $\widehat{\lambda}$ agree:

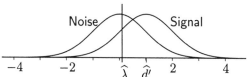

Numerically, the distance from μ_n up to $\widehat{\lambda}$ equals 0.10 and that from $\widehat{\lambda}$ up to $\widehat{d'}$ equals 0.92, so the center of the signal distribution lies $0.10 + 0.92 = 1.02$ units above the center of the noise distribution. Algebraically, the combination of steps 1 and 2 is

$$\widehat{d'} = \widehat{\lambda} - (\widehat{\lambda} - \widehat{d'})$$

$$= 0.10 - (-0.92) = 1.02.$$

From these calculations, detection performance in the first session is described by the parameter values $\widehat{d'} = 1.02$ and $\widehat{\lambda} = 0.10$.

To work in general, one must write these steps as equations. Because $Z(p)$ is the inverse of the cumulative Gaussian distribution, that is, the point for which $p = \Phi(z)$, the first step says that

$$Z(1-f) = \widehat{\lambda}.$$

The symmetry of the Gaussian distribution means that $Z(1-f) = -Z(f)$ (draw a picture), and that a simpler way to relate the false-alarm rate to the criterion is

$$\widehat{\lambda} = -Z(f). \tag{2.3}$$

Using the same two steps, the relationship between the hit rate and the estimates is

$$Z(1-h) = \widehat{\lambda} - \widehat{d'},$$

$$Z(h) = \widehat{d'} - \widehat{\lambda}.$$

Now use Equation 2.3 to replace $\widehat{\lambda}$ by $-Z(f)$ and solve for $\widehat{d'}$ to get

$$\widehat{d'} = Z(h) - Z(f). \tag{2.4}$$

When using Equations 2.3 and 2.4, it is helpful to verify that the criteria have been placed correctly and the right areas have been used by drawing a picture of the distributions, as was done above. Without seeing a picture, it is easy to drop the sign from $Z(h)$ or get the order of the subtracted terms wrong. Errors of this type are much easier to catch in a picture than in a formula.

Example 2.2: The results for the two sessions of the detection experiment described in Section 1.2 were

	h	f
First Session	0.82	0.46
Second Session	0.55	0.19

Show that the sessions differ in the criterion that was used more than in the detectability.

Solution: The signal-detection parameters for the first section were calculated in the three steps above. For the second session, Equations 2.3 and 2.4 give

$$\widehat{\lambda} = -Z(f) = -Z(0.19) = 0.88,$$

$$\widehat{d'} = Z(h) - Z(f) = Z(0.55) - Z(0.19) = 0.12 - (-0.88) = 1.00.$$

Summarizing these calculations,

	$\widehat{d'}$	$\widehat{\lambda}$
First Session	1.02	0.10
Second Session	1.00	0.88

It is immediately clear from this table that the criterion was shifted by the change in instructions, but the detectability of the signal remained the same. By converting from h and f to $\widehat{d'}$ and $\widehat{\lambda}$, the stability of detection, which was only suggested by the original proportions, becomes quantitatively clear. A formal statistical test of these conclusions, using the sampling model that will be developed in Chapter 11, is given in Example 11.6.

The two pairs of observations, (h, f) and $(\widehat{d'}, \widehat{\lambda})$, give different ways to view the outcome of a detection experiment. Each pair of numbers fully describes the results of the experiment. Although one pair can be converted to the other, each pair conveys information that is not so immediately obtained from the other pair. The signal-detection analysis translates the information provided by h and f into quantities that are much more descriptive of the detection process. Because the estimates $\widehat{d'}$ and $\widehat{\lambda}$ derive from the theoretical parameters of the signal-detection model, they separate the detection and the decision aspects of the observer's behavior in a way that h and f do not. Both types of information—the hit and false-alarm rates and the detection statistics—should be reported when describing the results of a detection task.

A problem sometime arises when one attempts to estimate detection parameters from a small sample of data. Although there is always some chance that the observer makes an error under the Gaussian detection model, with a small number of trials these events may never happen. The observer may respond YES on all the signal trials or NO on all the noise trials. These zero frequencies make either $h = 1$ or $f = 0$, so either $Z(h)$ or $Z(f)$ is undefined. Estimation of d' using Equation 2.4, which requires these values, does not work. Figure 2.2 illustrates the difficulty. The distributions are widely spaced, and either almost all of the signal distribution falls above the criterion or almost all of the noise distribution falls below it. However, unless some observations from each distribution fall on both sides of the criterion, it is impossible to tell how far apart to place the means. The distributions could be as shown in Figure 2.2, but they could as well lie somewhat closer or much farther apart.

How to deal with such data poses a dilemma. One possibility is to arbitrarily assign a small value to the empty category and proceed normally.

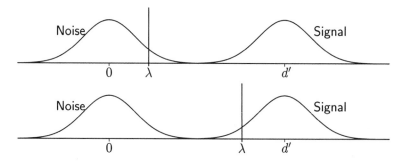

Figure 2.2: Distributions and criteria that represent data for which $h = 1$ (upper panel) or $f = 0$ (lower panel).

For example, a frequency of 1 observations, or $\frac{1}{2}$ observation, or $\frac{1}{10}$ observation might be assigned to an otherwise empty false-alarm category. If the noise stimulus has been presented on N_n trials, these values correspond to values of f of $1/(N_n + 1)$, $1/(2N_n + 1)$, or $1/(10N_n + 1)$, respectively. Giving a value to f lets Equations 2.3 and 2.4 be used to estimate λ and d'. Similarly, when $h = 1$, one can deduct 1, $\frac{1}{2}$, or $\frac{1}{10}$ observation from the number of hits to allow estimates to be made. The difficulty with these solutions is that even though the value that is substituted is arbitrary, it has an appreciable effect on the estimate. For example, suppose that in some study every one of 100 noise trials were correctly identified and 80 of 100 signal trials were hits. The three alternatives for the false-alarm rate give estimates of d' of 3.17, 3.42, and 3.93. No one of these numbers is more correct than another. It is important to recognize the arbitrariness of these values and to avoid drawing any conclusions that depend on the specific substitution.

An alternative to assigning a specific value to f or h is simply to acknowledge the indeterminateness and to treat the condition as implying a "large" detectability of the signal, without attempting to assign it an exact numerical value. In particular, the unknown estimate is larger than that of any comparable condition in which all the frequencies are positive. This problem will be revisited in Section 4.6.

2.4   Measuring bias

Although the parameter λ is the most direct way to express the placement of the observer's criterion, it is not the best way to measure the bias. Its value depends on the false-alarm rate, but not on the hit rate. Yet, how one interprets a particular criterion needs to take the detectability of the

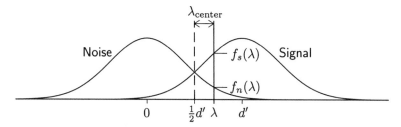

Figure 2.3: Measures of bias. The centered criterion λ_{center} is measured from the point between the signal and noise means. The heights of the density functions used to calculate the likelihood ratio are marked.

signal into account. For example, when $d' = 0.2$, the criterion $\lambda = 0.5$ represents a bias toward NO responses, but when $d' = 2.0$, it represents a bias toward YES responses—draw pictures of the distributions to see how the situations differ. A better descriptive measure of the bias takes account of the distributions of both signal and noise. There are several ways to combine both distributions in a bias measure, two of which are considered here.

One way to create a better measure of the bias is to express the position of the criterion relative to a point halfway between the signal and the noise distribution (Figure 2.3). A value of 0 for this measure indicates that the criterion is placed exactly between the distributions, implying that there is no preference for YES or NO responses and that the probabilities of misses and false alarms are the same. A negative displacement of the criterion relative to the center puts it nearer to the noise distribution and indicates a bias to say YES, and a positive value indicates a criterion nearer to the signal distribution and indicates a bias to say NO. The point between the distributions is $d'/2$ units above the mean of the noise distribution, so the placement of the criterion relative to this center is found by subtracting this midpoint from λ. Use the subscript $_{center}$ to distinguish the criterion measured this way from when it is measured from the noise mean:[2]

$$\lambda_{center} = \lambda - \tfrac{1}{2}d'. \tag{2.5}$$

Both λ and λ_{center} refer to the same criterion; they differ in the origin from which this criterion is measured. The use of a centered criterion is particularly natural when the origin of the decision axis is placed between the distributions, as described at the end of Section 2.1.

An estimate of λ_{center} can be found from $\hat{d}'$ and $\hat{\lambda}$, or it can be found directly from f and h. Substituting the estimates of λ and d' from Equations

[2]Macmillan and Creelman (1991) denote the measure λ_{center} by c.

2.3 and 2.4 into Equation 2.5 gives an estimate in terms of the response probabilities:

$$\widehat{\lambda}_{center} = \widehat{\lambda} - \tfrac{1}{2}\widehat{d'} = -\tfrac{1}{2}[Z(f) + Z(h)]. \tag{2.6}$$

A second measure of the bias expresses the propensity to say YES or NO by the relative heights of the two distribution functions at the criterion. Denote this ratio by β (the lowercase Greek letter beta):

$$\beta = \frac{f_s(\lambda)}{f_n(\lambda)}. \tag{2.7}$$

The two heights are marked in Figure 2.3 on page 27. In this figure, $\lambda = 1.80$ and $d' = 2.55$, and the height of the density functions at λ (using Equation A.43) are

$$f_n(\lambda) = \varphi(\lambda) = \frac{1}{\sqrt{2\pi}}e^{-\lambda^2/2} = \frac{1}{\sqrt{2\pi}}e^{-1.80^2/2} = 0.79,$$

and $f_s(\lambda) = \varphi(\lambda - d') = \varphi(-0.75) = 0.301$. The bias ratio is

$$\beta = \frac{f_s(\lambda)}{f_n(\lambda)} = \frac{\varphi(\lambda - d')}{\varphi(\lambda)} = \frac{\varphi(-0.75)}{\varphi(1.80)} = \frac{0.301}{0.079} = 3.81.$$

The ratio β is an instance of a *likelihood ratio*, which will play an important role in the general development of detection theory in Chapter 9.

To understand this likelihood-ratio bias measure, it helps to see how it varies with the position of the criterion. Figure 2.4 shows the two distributions of the equal-variance model and the ratio of their heights. Directly between the two means, the distributions are of equal height and the ratio is one; to the left of this center, the noise distribution is higher and the ratio is less than one; to the right of it, the signal distribution is higher and the ratio is greater than one.

Figure 2.5 shows noise and signal distributions with the criteria that might be adopted by three different observers. The observer with the criterion λ_1 is most biased toward saying YES. That observer makes both many hits and many false alarms. For this observer, $f_n(\lambda_1) > f_s(\lambda_1)$ and β is less than one. The observer with criterion λ_3 is just the opposite. Here the bias is such as to make few false alarms and, in consequence, a lower hit rate than in the other condition. Now $f_n(\lambda_3) < f_s(\lambda_3)$ and β exceeds one. The criterion λ_2 has an intermediate placement, for which $f_n(\lambda_2) = f_s(\lambda_2)$ and $\beta = 1$.

The major defect with β as a bias measure is that it is asymmetrical. As Figure 2.4 shows, bias favoring YES responses is indicated by values in the

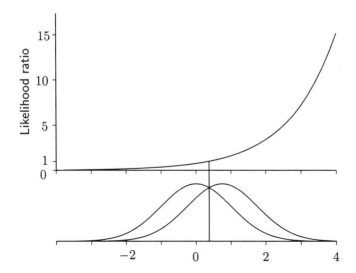

Figure 2.4: The ratio of the heights of two unit-variance Gaussian density functions. The means of the distributions differ by $d' = 0.75$. The vertical line marks the point where both distributions have equal height.

narrow range between 0 and 1, while bias favoring NO responses is indicated by values ranging from 1 all the way to ∞. Much more room is available on the NO side than on the YES side. This asymmetry is removed by taking the natural logarithm[3] of the likelihood ratio:

$$\log \beta = \log \left[\frac{f_s(\lambda)}{f_n(\lambda)} \right] = \log f_s(\lambda) - \log f_n(\lambda). \qquad (2.8)$$

The placement of the criterion in Figure 2.3 gives $\log \beta = \log(3.81) = 1.34$.

For the equal-variance Gaussian model, the distribution functions can be substituted into Equation 2.8 to express $\log \beta$ in terms of the other parameters. Substituting the Gaussian density functions with the appropriate means and variances (Equation A.42) gives

$$\log \beta = \log \left\{ \frac{1}{\sqrt{2\pi}} \exp\left[-\tfrac{1}{2}(\lambda - d')^2 \right] \right\} - \log \left\{ \frac{1}{\sqrt{2\pi}} \exp\left[-\tfrac{1}{2}\lambda^2 \right] \right\}$$

---

[3]Here and throughout this book the expression $\log(x)$ refers to the natural logarithm, that is, to the logarithm to the base $e = 2.71828$. This type of logarithm is sometimes denoted by $\ln(x)$, particularly on calculators. Common logarithms (to the base 10) are not used in this book. One reason for the importance of the natural logarithm is that it is the inverse of exponentiation. You can undo one by doing the other: $\log[\exp(x)] = x$ and $\exp[\log(x)] = x$, as in derivations such as that leading to Equation 2.9 below.

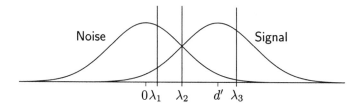

Figure 2.5: Distributions of signal and noise with three decision criteria.

$$= \log \frac{1}{\sqrt{2\pi}} + \log\left\{\exp\left[-\tfrac{1}{2}(\lambda - d')^2\right]\right\} - \log \frac{1}{\sqrt{2\pi}} - \log\left\{\exp\left[-\tfrac{1}{2}\lambda^2\right]\right\}$$

$$= -\tfrac{1}{2}(\lambda - d')^2 + \tfrac{1}{2}\lambda^2$$

$$= d'(\lambda - \tfrac{1}{2}d'). \tag{2.9}$$

The value of $\log\beta$ is zero at the point where the density functions cross, which for this model occurs when the criterion is at $d'/2$, exactly between the distributions. It is negative for criteria lying to left of this midpoint, favoring YES responses, and positive for criteria lying to the right, favoring NO responses. As the form of Equation 2.9 shows, the log likelihood ratio is a linear function of λ, so would be a straight line if plotted on Figure 2.4.

An estimate of the bias is found by substituting estimates of d' and λ into Equation 2.9:

$$\widehat{\log\beta} = \hat{d}'(\hat{\lambda} - \tfrac{1}{2}\hat{d}'). \tag{2.10}$$

Substituting the estimates $\hat{d}'$ and $\hat{\lambda}$ (Equations 2.3 and 2.4) into this formula shows that $\widehat{\log\beta}$ is half the difference of the squares of the Z transforms of f and h:[4]

$$\widehat{\log\beta} = \tfrac{1}{2}[Z^2(f) - Z^2(h)]. \tag{2.11}$$

The two measures λ_{center} and $\log\beta$ are quite similar, as a comparison of Equations 2.5 and 2.9 shows. They differ in that $\log\beta$ is scaled by the detectability relative to λ_{center}:

$$\log\beta = d'\lambda_{\text{center}} \qquad \text{and} \qquad \lambda_{\text{center}} = \log\beta/d'. \tag{2.12}$$

Both measures are satisfactory indicators of response bias, and it is hard to choose one over the other. When comparing the bias in conditions in which the signals are equally detectable, the choice is inconsequential; as Equations 2.12 indicate, both measures order the conditions in the same way.

---

[4]A squared function $f^2(x)$ is a shorthand for the square of the function, $[f(x)]^2$.

However, when the conditions differ in detectability, the two measures can lead to different conclusions. With large differences in d', the measures can even order two conditions differently. It would be nice to be able to choose either $\log \beta$ or λ_{center} based on some fundamental property of the signal-detection model. The choice would be based on a clear understanding of how the observer reached the decision. However, theories of the observer's performance are not sufficiently developed to resolve the issue. It is likely that real behavior is sufficiently variable and influenced by experimental conditions to preclude a single choice. In the end, it is risky to make quantitative comparisons of bias across conditions with widely varying detection parameters.

This book emphasizes the likelihood-ratio measure $\log \beta$ for two plausible, but not definitive, reasons. First, the measure generalizes more easily than λ_{center} to situations in which the distributions of signal and noise events have different variance, are non-Gaussian, or are multivariate. Second, the likelihood-ratio definition seems somewhat closer to what is intended by the word "bias." An observer who is biased toward a particular alternative will choose it even when the evidence on which the choice was made is more likely to have occurred were the other alternative true. This sense of bias is more similar to a likelihood ratio than a criterion placement. In what follows the unqualified term *bias* refers to $\log \beta$.

Example 2.3: Estimate the bias of the observer in the two sessions in the example of Section 1.2 and Example 2.2.

Solution: Starting with the fitted detection model for the first of the two sessions and using Equations 2.10 and 2.6 gives the estimates

$$\widehat{\log \beta} = \widehat{d'}(\widehat{\lambda} - \tfrac{1}{2}\widehat{d'}) = 1.02(0.10 - 1.02/2) = -0.42,$$

$$\widehat{\lambda}_{\text{center}} = \widehat{\lambda} - \tfrac{1}{2}\widehat{d'} = 0.10/1.02/2 = -0.41.$$

The calculation can also start with the raw detection results and use Equations 2.11 and 2.6 to find the bias. For the second session:

$$\widehat{\log \beta} = \tfrac{1}{2}[Z^2(f) - Z^2(h)] = \tfrac{1}{2}[Z^2(0.19) - Z^2(0.55)]$$

$$= \tfrac{1}{2}[(-0.878)^2 - (0.126)^2] = 0.38,$$

$$\widehat{\lambda}_{\text{center}} = -\tfrac{1}{2}[Z(f) + Z(h)] = \tfrac{1}{2}[(-0.878) + 0.126] = 0.38.$$

The two measures $\widehat{\log \beta}$ and $\widehat{\lambda}_{\text{center}}$ are almost identical here because d' is almost exactly unity. Moreover, the fact that d' is about the same in the two sessions means that either $\widehat{\log \beta}$ or $\widehat{\lambda}_{\text{center}}$ leads to the same conclusion about the relationship between the bias of the two sessions.

2.5 Ideal observers and optimal performance

One reason to construct a theoretical detection model is to determine how well the best possible observer could do. Such a hypothetical individual, who is able to make optimal use of the information available from the stimulus, is known as an *ideal observer*. Like a real observer, an ideal observer is limited by the intrinsic uncertainty of the stimulus events, so cannot attain perfect performance. The performance of the ideal observer indicates what is possible, given the limits imposed by the random character of the signals.

The ideal-observer model is not a description of real data. Real observers are not ideal, and they usually fall short of ideal behavior. The deviations between idea and real performance are interesting, because they tell something about the actual perceptual processes or decision-making behavior. The ideal observer gives the standard needed to make this comparison.

In complex tasks there can be several sorts of ideal observers, each optimally using different information about the stimulus. For example, an observer who knows exactly the form of the distribution of the signal and noise events can perform differently than an observer who knows less about the shape of these distributions. This section considers an observer in a yes/no detection task who is aware of the statistical properties of the two types of stimulus but has no specific control over these characteristics. This description is a good ideal representation for an observer who can modulate performance through changes in the decision criterion but who cannot control the nature of what is observed.

Specifically, consider an observer who has access to the stimulus only through the random variables X_n and X_s. This observer knows the forms of these distributions and the proportion of the trials on which a signal occurs. This observer can modulate the decision criterion λ (or select the likelihood ratio at the decision criterion) so that the largest possible number of correct responses are made. The signal-detection model lets this optimal value be determined.

The first step in determining the optimal criterion is to calculate the correct-response probability P_C as a function of the decision criterion λ. The observer can then choose a criterion λ^* that maximizes P_C with respect to λ. The first part of the calculation is very general. The probability of a correct response is the sum of the probability of a hit and the probability of a correct rejection. The probability of a signal trial is s and of a noise trial is $1 - s$, and the probabilities of correct responses to signal and noise trials are $1 - F_s(\lambda)$ and $F_n(\lambda)$, respectively. Combining these probabilities, the overall probability of a correct response is

$$P_C = P(\text{signal trial and YES}) + P(\text{noise trial and NO})$$

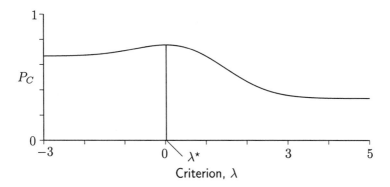

Figure 2.6: *The probability of a correct response as a function of the criterion for signals with* $d' = 1.2$ *occurring on* $s = \frac{2}{3}$ *of the trials. The value of* P_C *is greatest when the criterion is placed at* λ^*.

$$= P(\text{signal trial})P(\text{YES}|\text{signal}) + P(\text{noise trial})P(\text{NO}|\text{noise})$$

$$= s[1 - F_s(\lambda)] + (1 - s)F_n(\lambda). \tag{2.13}$$

Figure 2.6 shows this function for the equal-variance Gaussian model in a study with $s = \frac{2}{3}$ signal trials. On the left, when all responses are YES, $P_C = s = \frac{2}{3}$, and on the right, when all responses are NO, $P_C = 1 - s = \frac{1}{3}$. In between, P_C rises to a maximum, then declines. The best performance is obtained by setting the criterion to hit this maximum point.

The value λ^* of the criterion for which P_C is greatest is found using calculus.[5] Differentiating Equation 2.13 with respect to λ gives

$$\frac{dP_C}{d\lambda} = -s\frac{dF_s(\lambda)}{d\lambda} + (1 - s)\frac{dF_n(\lambda)}{d\lambda}$$
$$= -sf_s(\lambda) + (1 - s)f_n(\lambda).$$

By setting this quantity to zero and doing some algebraic manipulation, one finds that at the optimal criterion λ^*, the likelihood ratio β is equal to the ratio of the probability of a noise trial to the probability of a signal trial, a quantity known as the odds (page 228):

$$\beta^\star = \frac{f_s(\lambda^\star)}{f_n(\lambda^\star)} = \frac{1 - s}{s}. \tag{2.14}$$

---

[5] The extremum (minimum or maximum) of a smooth function occurs at the point where its derivative (or slope) is zero—the hill in Figure 2.6 is flat at the very top. Readers who are unfamiliar with the calculus can skip to the answer in Equation 2.14.

The optimal bias is minus the logarithm of this ratio, which is known as the logit (page 228):

$$\log \beta^{\star} = \log \frac{1-s}{s} = -\text{logit}(s). \tag{2.15}$$

This result will be developed from a different background in Section 9.2.

When signals and noise events are equally likely (i.e., when $s = \frac{1}{2}$), Equation 2.14 says that the criterion should be chosen so that the two densities are equal, that is, so that $f_n(\lambda^{\star}) = f_s(\lambda^{\star})$. More simply, the optimal criterion lies at the point where the two distribution functions cross.[6] In Figure 2.5 the criterion λ_2 is located at this optimal point. Positioning the optimal criterion at the crossover makes good intuitive sense. To the left of the crossover, $f_n(x)$ is greater than $f_s(x)$, so as one moves the criterion in this direction, false alarms are being added more rapidly than hits are being increased, and P_C goes down. To the right of the crossover, $f_n(x) < f_s(x)$, so as one moves λ in that direction, hits decrease more rapidly than do false alarms, and P_C also goes down.

Unequally frequent signal and noise events displace the optimal point away from the crossover. When signals are more likely than noise events, the odds $s/(1-s)$ are larger than one, and the optimal criterion is at a point where $f_n(\lambda^{\star}) > f_s(\lambda^{\star})$. This point lies to the left of the place where the distributions cross. When signals are rare, the optimal point shifts to the right, to a position at which evidence favoring noise events is given greater weight, so that $f_n(\lambda^{\star}) < f_s(\lambda^{\star})$.

The analysis thus far has not used any information about the actual distribution of the variables X_n and X_s. Results such as Equations 2.14 and 2.15 apply to any pair of distributions. When the forms of the distribution functions are known, the position of the criterion can be worked out exactly. Under the equal-variance Gaussian model, Equation 2.9 gives the relationship among d', the optimal criterion $\lambda^{\star}$, and the optimal bias $\log \beta^{\star}$:

$$\log \beta^{\star} = d'(\lambda^{\star} - \tfrac{1}{2}d').$$

Substituting the optimal bias from Equation 2.15 and solving for optimal criterion gives

$$\lambda^{\star} = \tfrac{1}{2}d' - \frac{\text{logit}(s)}{d'}. \tag{2.16}$$

---

[6]There are distributions for which Equation 2.14 has several solutions or where the solution to Equation 2.14 minimizes P_C. However, these configurations do not pose practical problems. When there are several solutions, the appropriate one is usually clear. Solutions that minimize P_C are primarily curiosities. More rigorous optimality criteria can be established when needed.

The point $d'/2$ is halfway between the centers of the distributions and is the optimal point when $s = \frac{1}{2}$. The second term adjusts the optimal criterion for unequal stimulus frequencies, moving it downward when $s > \frac{1}{2}$ and upward when $s < \frac{1}{2}$. Expressed in terms of the centered criterion λ_{center}, the picture is even simpler. The optimal centered criterion is zero when $s = \frac{1}{2}$, and is shifted to $-\text{logit}(s)/d'$ when the frequencies are unequal.

Example 2.4: Suppose that in a difficult detection task $d' = 1.2$ and that signals occur on two thirds of the trials. Where should the criterion be placed under the equal-variance Gaussian model? What is the maximum probability of a correct response that an observer can attain? How do these results change when only one trial in 10 is a signal?

Solution: The logit of the signal probability is

$$\text{logit}(s) = \log \frac{\frac{2}{3}}{1 - \frac{2}{3}} = \log 2 = 0.693.$$

By Equation 2.16, the optimal criterion is

$$\lambda^\star = \frac{1}{2}d' - \frac{\text{logit}(s)}{d'} = \frac{1.2}{2} - \frac{0.693}{1.2} = 0.600 - 0.5781 = 0.022.$$

The probability of a correct response using this optimal criterion (Equation 2.13) is

$$P_C = s[1 - \Phi(\lambda - d')] + (1 - s)\Phi(\lambda)$$

$$= \frac{2}{3}[1 - \Phi(0.022 - 1.200)] + \frac{1}{3}\Phi(0.022)$$

$$= \frac{2}{3}(1 - 0.120) + \frac{1}{3}(0.509) = 0.757.$$

Figure 2.6 was drawn using this set of parameters. When signals are rare, with $s = \frac{1}{10}$, the log-odds are

$$\text{logit}(s) = \log \frac{\frac{1}{10}}{\frac{9}{10}} = -\log 9 = -2.197,$$

and the optimal criterion is

$$\lambda^\star = \frac{1.2}{2} - \frac{-2.197}{1.2} = 0.600 + 1.831 = 2.431.$$

This criterion actually lies on the far side of the mean of the signal distribution. The bias to say NO is very strong. Notwithstanding this asymmetry, the probability of a correct response is substantial:

$$P_C = \frac{1}{10}[1 - \Phi(2.431 - 1.200)] + \frac{9}{10}\Phi(2.431) = 0.904.$$

This large probability reflects the fact that even before the trial starts, the observer can be fairly sure that no signal will occur. Only the strongest signals receive a YES response, and P_C is little greater than the value of 0.9 that could be obtained by unconditionally saying NO.

The probability P_C of a correct response is only one of the many quantities that an observer may chose to maximize. Where there are costs or values associated with the alternatives, the observer can adjust the criterion to minimize the cost or maximize the gain. If false alarms are expensive and misses are cheap, then an optimal observer will shift λ to reduce the false-alarm rate. Specifically, suppose the value of the various outcomes are given by $V(\text{hit})$, $V(\text{miss})$, and so forth—the first value is usually positive and the second negative. The average value of a trial is

$$
\begin{aligned}
\mathsf{E}(V) = {} & P(\text{signal and YES})V(\text{hit}) + P(\text{signal and NO})V(\text{miss}) \\
& + P(\text{noise and YES})V(\text{false alarm}) \\
& + P(\text{noise and NO})V(\text{correct rejection}).
\end{aligned}
\tag{2.17}
$$

This equation can be expanded in the manner of Equation 2.13 and its maximum value found. It turns out the optimal criterion can be expressed in terms of the difference between the value of a correct response and the value of an error for each of the stimulus types, which are, in effect, the costs of the two types of errors:

$$
c_m = V(\text{hit}) - V(\text{miss}),
$$

$$
c_f = V(\text{correct rejection}) - V(\text{false alarm}).
$$

The result (derived in Problem 2.8) is to add a second term to the optimal bias that was found in Equation 2.15:

$$
\log \beta^\star = \log \frac{c_f}{c_m} - \text{logit}(s).
\tag{2.18}
$$

Real observers make similar accommodations, albeit not always to the optimal point. An experimenter can easily manipulate an observer's criterion by changing the payoff for the various responses. It is instructive to consider how the real or perceived costs of a decision alter the bias in real-world detection tasks such as those mentioned at the start of Chapter 1.

Reference notes

The sources cited in the previous chapter apply here as well. Greater detail on some of the alternative approaches (logistic distributions, other bias measures, etc.) are given by Macmillan and Creelman (1991), who also summarize some empirical evidence. The choice theory mentioned in connection with the logistic distribution was originally formulated by Luce (1959) and is described in Atkinson, Bower, and Crothers (1965).

Exercises

2.1. Calculate P_H and P_F for the five-parameter detection model of Section 2.1 with $X_n \sim \mathcal{N}(2,4)$, $X_s \sim \mathcal{N}(7,9)$, and a criterion at 5. Show by calculation that the same values are obtained from the standard three-parameter model with $X_s \sim \mathcal{N}(2.5, 2.25)$ and $\lambda = 1.5$.

2.2. Suppose that, in the signal-detection model for a simple detection task, the means of the signal distribution and the noise distribution differ by four fifths of the standard deviation of the noise distribution and that the variance of the signal distribution is three times that of the noise distribution. What are the hit rate and the false-alarm rate for a decision criterion placed one half standard deviation of the noise distribution above the mean of the noise distribution? Start by drawing a picture, then do the calculation.

2.3. Suppose that an observer sets $\lambda = 0.8$ in a detection task with $d' = 1.4$. What are the predicted hit rate and the false-alarm rate?

2.4. What are β and $\log \beta$ in Problem 2.2?

2.5. Draw pictures illustrating the situation described in the first paragraph of Section 2.4. What values of $\log \beta$ are implied in the two cases?

2.6. In a detection experiment 1000 trials are run, 600 with the signal and 400 with noise alone. The resulting responses are

	NO	YES	
Noise	327	73	400
Signal	104	496	600

Estimate d', λ, $\log \beta$, and λ_{center}. Draw a picture illustrating the model.

2.7. When the distributions are as estimated in Problem 2.6, what value of λ produces the most correct responses? How close is the observed probability of a correct response to its optimal value?

2.8. Suppose that costs are assigned to the two types of errors, with false alarms being charged a cost of c_f and misses being charged a cost of c_m.

a. Expand Equation 2.17 and insert probabilities to find the average cost associated with a criterion placed at position λ.

b. Maximize this expression and show that the optimal bias is given by Equation 2.18.

2.9. Consider an observer watching for a rare event, such as one that occurs on only one percent of the trials. The event by itself is reasonably detectable, with $d' = 2$. What would an observer minimizing errors do here? Using this criterion, what is the proportion of signals that are overlooked? Now suppose that this omission probability is too large. To induce the observer to reduce omissions, a reward of $50 is given for each correct detection and $1 for each correct rejection. How does the observer's performance change if it is adjusted to maximize monetary return? How does this change affect the false-alarm rate?

2.10. The theory of signal detectability has been extensively applied in radiology. Consider someone examining X-ray films for signs of a rare disorder. Discuss the various tradeoffs in performance and outcome inherent to the task, relating the performance to the situation described in Problem 2.9.

Chapter 3

Operating characteristics and the Gaussian model

In Section 1.2, data from two detection sessions with different biasing instructions were described. The analysis of these data showed that the sessions differed in the observer's bias toward YES or NO responses, but not appreciably in the detectability of the signal. In these analyses, each session's data were treated independently. A more comprehensive analysis should tie the two sets of observations together in a single model of the detection task. This chapter describes this integration. It also shows how to represent data from several detection conditions in a convenient, instructive, and widely used graphical form. This representation lets one investigate the adequacy of the assumptions that underlie the Gaussian signal-detection model and fit a model in which the variances of the signal and noise distributions are unequal.

3.1 The operating characteristic

Consider the results from the two sessions of Section 1.2. Fitting an equal-variance Gaussian model to these data, as in Examples 2.2 and 2.3, gives the parameter estimates at the top of Figure 3.1. As these numbers show, the estimates of d' are very similar, but those of λ and $\log \beta$ differ considerably. These values suggest that the proper representation of the full set of observations should use the same pair of distributions in the two sessions, but change the criterion. This representation is illustrated at the bottom of Figure 3.1. Using the same distributions for both sessions implies that the characteristics of the stimuli (which are controlled by the experimenter)

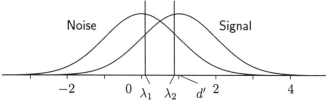

	h	f	$\widehat{d'}$	$\widehat{\lambda}$	$\widehat{\log \beta}$
First Session	0.82	0.46	1.02	0.10	−0.42
Second Session	0.55	0.19	1.00	0.88	0.38

Figure 3.1: *Data and parameter estimates for the two detection sessions of Section 1.2 and a representation using stimulus distributions separated by $d' = 1.01$ with specific criteria at $\lambda_1 = 0.10$ and $\lambda_2 = 0.88$.*

do not change, while using session-specific criteria implies that the decision rules applied to these distributions are different.

The change in instructions that shifted the criteria in Figure 3.1 produced a trade-off between the two types of correct response, hits and correct rejections. The higher criterion produced many correct rejections but few hits, while a low criterion increased the hit rate at the cost of the correct-rejection rate. Put another way, changes in criterion cause the hit rate and the false-alarm rate to vary together. The trade-off is illustrated more directly by plotting the hit rate against the false-alarm rate, in the manner of a scatterplot, as shown in Figure 3.2. The false-alarm rate is plotted on the abscissa (horizontal axis) and the hit rate on the ordinate (vertical axis). Each session's results are represented by a point, S_1 for the first session and S_2 for the second session. Because the hit rate exceeds the false-alarm rate, these points lie above the diagonal of the square that connects the point $(0, 0)$ to the point $(1, 1)$. The outcome of any detection study corresponds to a point in this square, generally in the upper triangle.

The theoretical counterpart of Figure 3.2 plots the probabilities P_F and P_H generated by a particular signal-detection model. This plot is particularly valuable in illustrating how constraints on the parameters limit the predictions. Consider what is possible for an observer working according to the Gaussian model who can vary the criterion but who cannot alter d'. Figure 3.3 shows four points from such an observer, with the corresponding distributions in insets. Each point is based on the Gaussian model with $d' = 1.15$—note that the relative positions of the distributions do not change. However, the criteria differ. Bias toward NO responses gives the point at the lower left, and bias toward YES responses gives the point at the upper right.

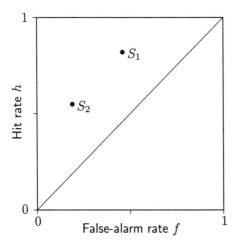

Figure 3.2: A plot of the false-alarm rate f and the hit rate h for the two detection sessions of Section 1.2.

The full range of performance available to this observer is shown by a line on this graph. When λ is very small, both P_F and P_H are large, leading to a point near the (1, 1) corner of the graph. As λ is increased, the area above the criterion decreases, as do P_F and P_H, and the point (P_F, P_H) traces out the curve shown in the figure. Eventually the prediction approaches (0, 0) when λ is so large that both distributions fall almost completely below it.

The solid line in Figure 3.3, which shows the possible range of performance as the bias is adjusted, is known as an *operating characteristic*. The original application of the statistical decision model to detection problems grew out of work on radio reception, where these curves were known as receiver operating characteristics. This name has stuck; the curve is often called a *receiver operating characteristic* or, more briefly, a *ROC curve*. Because it shows the performance possible for a fixed degree of discriminability, that is, a fixed sensitivity to the signal, the curve is also known as an *isosensitivity contour*. Both terms are used in this book.

Figure 3.3 shows the operating characteristic for a signal with detectability $d' = 1.15$. By varying the detectability of the signal—say, by changing its intensity—one obtains a family of curves. Figure 3.4 shows several members of this family. These are drawn for the equal-variance Gaussian model (using the methods described in the next section) and are labeled with values of d'. When $d' = 0$ the hit rate and the false-alarm rates are identical and the operating characteristic lies along the diagonal of the square. For positive values of d', the curves lie above this diagonal, moving close and closer to the upper left-hand corner as d' increases. When d' is

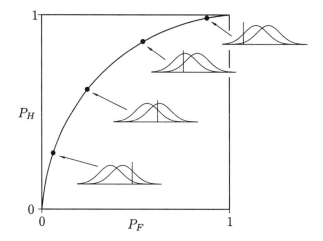

Figure 3.3: *The operating characteristic or isosensitivity contour for a detection task derived from the equal-variance Gaussian model with* $d' = 1.15$. *The inset diagrams show the distributions of the random variables* X_n *and* X_s *and the criterion* λ *at four typical points.*

very large, the operating characteristic is almost identical to the two lines at right angles that make up the left and top sides of the square. This operating characteristic describes performance when the signal is so strong that the observer can almost completely avoid making errors.

In an important sense, the operating characteristics are what describe the sensitivity of an observer to a particular signal. Together, the strength of that signal and the receptivity of the observer determine how detectable that signal will be. The combination selects the operating characteristics associated with that d', and the outcome of any study falls on that line. The point on that line that is actually observed is determined by the observer's bias.

3.2   Isocriterion and isobias contours

Aspects of the signal-detection model other than the bias can be varied to produce contours in the (P_F, P_H) space. *Isocriterion contours* are created by holding λ constant and varying the sensitivity. Because the criterion and the false-alarm rate determine each other (remember that P_F is the probability that $X_n > \lambda$), the isocriterion contours are vertical lines at constant values of P_F (Figure 3.5, top). Contours such as these might describe the results of a study in which signals of different strengths were

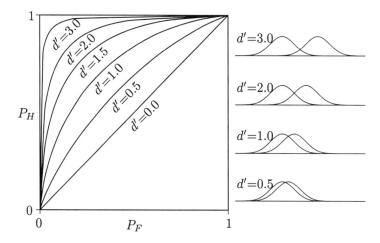

Figure 3.4: *The family of isosensitivity contours determined by different values of d' for the equal-variance Gaussian model. At the side are shown the noise and signal distributions for four of the contours.*

mixed with simple noise events, thus giving one false-alarm rate and several hit rates (one for each strength), with decisions based on a single criterion.

More interesting are the *isobias contours* created by holding one of the bias indices constant while varying the sensitivity. The lower two panels of Figure 3.5 show the lines of constant β or $\log \beta$ (left) and of constant (right). Both sets of isobias contours end in the upper left corner of the plot, a consequence of the fact that with sufficiently strong signals it is possible to attain near-perfect performance regardless of the bias. The curves otherwise have substantially different shapes, reflecting the different properties of the two measures. Each curve of constant likelihood ratio starts for very weak signals either at the point $(0, 0)$ (for positive values of $\log \beta$) or the point $(1, 1)$ (for negative values). In contrast, each curve of constant λ_{center} starts on a different point on the chance diagonal.

The isobias contours of either type cut across the isosensitivity contours of Figure 3.4. Thus, any point (P_F, P_H) in the space determines one level of sensitivity (a line from Figure 3.4) and one level of bias (a line from Figure 3.5). Similarly, any combination of sensitivity and bias determine, by the intersection of one line from each of the figures, a false-alarm rate and a hit rate.

Isocriterion and isobias curves are usually of less practical interest than are the isosensitivity curves. One reason is that it is easier to find situations in which one expects the sensitivity to remain constant while the bias changes than situations where one expects the criterion or the bias to remain constant while sensitivity changes. Another reason is that bias is a

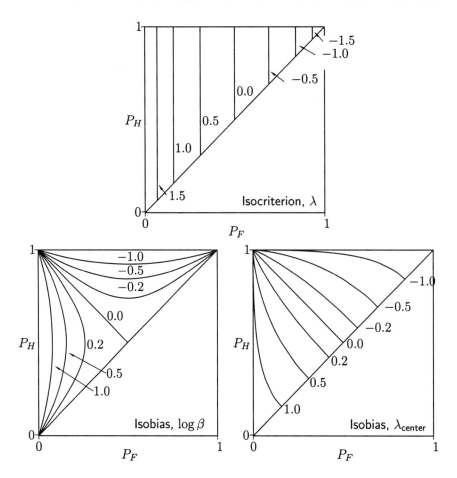

Figure 3.5: *Families of isocriterion curves (top) and isobias curves (bottom) for the equal-variance Gaussian model. Isobias curves are shown for β or $\log \beta$ (left) and λ_{center} (right).*

more complex psychological construct than is sensitivity. It is easy to understand how the sensitivity of an observer to a particular physical stimulus could remain constant, but it is much harder to assert that bias, by any particular definition, should not vary as the signal changes. Finally, most research that uses detection theory is concerned with factors that influence detectability, not with those affecting the decision process. Differences in response bias in these studies are a nuisance to be removed, not a process to be studied. Determining on which isosensitivity curve the performance falls is most important.

3.3   The equal-variance Gaussian operating characteristic

The actual shape of the operating characteristics is derived theoretically from the distributions of the random variables X_s and X_n. First consider the process in its most general form. Suppose that $f_s(x)$ and $f_n(x)$ are the density functions of these random variables. The hit rate and the false-alarm rate are areas under these curves above the criterion point and are calculated as integrals of the corresponding density functions above those points:

$$P_F = \int_\lambda^\infty f_n(x)\,dx \quad \text{and} \quad P_H = \int_\lambda^\infty f_s(x)\,dx \tag{3.1}$$

(Equations 1.1 and 1.2). Denote these quantities, as functions of λ, by $P_F = F(\lambda)$ and $P_H = H(\lambda)$, respectively—here F and H stand for false alarms and hits, respectively, and are not cumulative distribution functions. If one could write these integrals as simple expressions, then one could solve $F(\lambda)$ for λ and substitute it in $H(\lambda)$ to get the isosensitivity curve. Working purely formally:[1]

$$\lambda = F^{-1}(P_F),$$

$$P_H = H(\lambda) = H[F^{-1}(P_F)]. \tag{3.2}$$

Although this procedure is simple in the abstract, a practical difficulty arises with the Gaussian distribution. For it, Equations 3.1 cannot be written as simple expressions, and they cannot be solved algebraically to give a formula for Equation 3.2. Gaussian operating characteristics, such as those in Figures 3.3 and 3.4 are constructed using tables or numerical methods.

The relationship between P_F and P_H implied by the operating characteristic is simplified by transforming the probabilities before they are plotted. The appropriate transformation here is the inverse Gaussian function $Z(p)$. After applying it, the operating characteristic is a straight line. In more detail, first recall (from Equations 2.2 on page 21) that the response probabilities under the Gaussian model with $\sigma_s^2 = 1$ are

$$P_F = 1 - \Phi(\lambda) = \Phi(-\lambda),$$

$$P_H = 1 - \Phi(\lambda - d') = \Phi(d' - \lambda).$$

---

[1] A function raised to a negative power is the inverse function: if $y = f(x)$, then $x = f^{-1}(y)$. This use of the exponent differs from the notation for positive powers, for which $f^2(x) = [f(x)]^2$ (see footnote 4 on page 30).

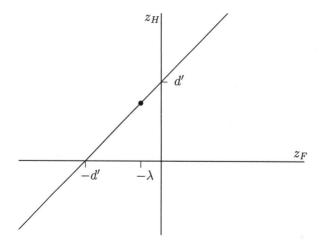

Figure 3.6: *The isosensitivity function for the equal-variance Gaussian model plotted in Gaussian coordinates. The solid point corresponds to performance with a criterion of* λ.

The formula for the hit rate takes the nonzero mean of X_s into account. Now transform these probabilities to $z_F = Z(P_F)$ and $z_H = Z(P_H)$. This transformation undoes the function $\Phi(z)$:

$$z_F = Z(P_F) = Z[\Phi(-\lambda)] = -\lambda,$$

$$z_H = Z(P_H) = Z[\Phi(d' - \lambda)] = d' - \lambda.$$

Eliminating λ from these equations gives

$$z_H = z_F + d'. \tag{3.3}$$

This equation describes the isosensitivity function when it is plotted on Gaussian transformed axes, that is, in *Gaussian coordinates*.

Figure 3.6 shows the operating characteristic plotted in these coordinates. Four things about it are important to notice:

- The function is linear. Plotted in Gaussian-tranformed coordinates, the isosensitivity function is a straight line. This fact is a consequence of the choice of the Gaussian form for X_n and X_s.
- The function has a 45° slope. This slope is a consequence of the equal-variance assumption. As will be seen in the next section, models with unequal variances have different slopes.
- The line crosses the axes at $-d'$ and d'. Thus, having drawn the line, one can easily read off the value of d' from either intercept or, having d', can easily draw the line.

Probabilities		Gaussian scores	
f	h	$Z(f)$	$Z(h)$
0.12	0.47	−1.18	−0.08
0.18	0.72	−0.92	0.58
0.20	0.58	−0.84	0.20
0.38	0.78	−0.31	0.77
0.51	0.77	0.02	0.74
0.66	0.92	0.41	1.40
0.77	0.96	0.74	1.75

Table 3.1: *Hit and false-alarm data obtained by varying the bias at without changing the signal.*

- Isocriterion contours are vertical lines at $-\lambda$. So the point corresponding to a model with a particular criterion λ lies at the point of the operating characteristic with a value of $-\lambda$ on the z_F axis.

Because the isosensitivity function in Gaussian coordinates is a straight line, the model is easy to apply to data from several bias conditions. First convert the observed hit and false-alarm proportions to Gaussian coordinates using either tables or a computer program. Then draw a straight line at 45° through these points. The intercept on the ordinate is an estimate of d'. This procedure is discussed in more detail in Section 3.5; for the moment an example will illustrate it.

Example 3.1: Table 3.1 shows the proportions that might be obtained from a seven-level manipulation of bias. Fit the equal-variance Gaussian model to these data and draw the operating characteristic.

Solution: The observed proportions f and h are plotted directly on the left in Figure 3.7. The points roughly trace out an isosensitivity curve, but there is sufficient scatter that an operating characteristic cannot be constructed by connecting them. The resulting line is neither smooth nor monotonic. The scatter is not surprising and is inevitable unless an extremely large number of observations have been obtained at each point, a condition fulfilled only by a few psychophysical experiments.

To fit the function the proportions are converted to Gaussian scores in the second part of Table 3.1 and plotted on the right in Figure 3.7. It is now easy to draw a line at 45° through the midst of them. The points are adequately fitted by this line—certainly they have no systematic curvature—and from the place where this line crosses the axes, $\widehat{d'}$

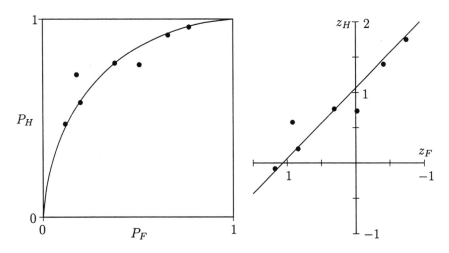

Figure 3.7: Plots of the data in Table 3.1 with fitted operating characteristics in probability coordinates (left) and Gaussian coordinates (right).

is apparently slightly greater than one. A more accurate estimate, obtained from a computer program, is $\widehat{d'} = 1.065$. To plot the isosensitivity function in its conventional form, points on the line in Gaussian coordinates are reconverted to probability coordinates using the transformation $p = \Phi(z)$ giving the theoretical line in the probability plot in Figure 3.7.

3.4   The unequal-variance Gaussian model

In the general Gaussian signal-detection model, the distributions of X_n and X_s can differ in their variance as well as their means. With the constraints placed on parameters of the noise distribution to give the model unique values (as discussed in Section 2.1), the distributions of the random variables associated with the two stimulus events are

$$X_n \sim \mathcal{N}(0,1) \qquad \text{and} \qquad X_s \sim \mathcal{N}(\mu_s, \sigma_s^2).$$

The resulting model is more flexible than the equal-variance version, and when data from several conditions are available, it often provides a superior fit. Figure 3.8 shows an example in which the signal distribution has greater variance than the noise distribution.

Unequal variability of the signal and noise events arises quite naturally in a number of situations. One explanation is based on the rules for the addition of random variables. Consider the detection of a pure tone of

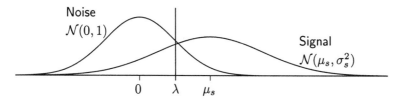

Figure 3.8: *Distributions of signal and noise under a Gaussian model in which the standard deviation of the signal distribution is 1½ times that of the noise distribution.*

known frequency ν in a white noise background. Suppose that the observer bases the detection on the output of a tuned detector responding to the intensity of the stimulation at the frequency ν. The "noise" here consists of the variation in the extent to which the white noise near the frequency ν excites the detector. This variation creates the variability of X_n. When the tone is added to the background, creating the signal-plus-noise condition, the background variability does not vanish. The added signal increases the response of the detector above that in the noise conditions, say, by an amount S:

$$X_s = X_n + S.$$

If S is a pure constant, without any variability itself, then the random variable X_s is simply a displaced version of X_n, and their variances are the same. However, if the signal has variability, then S is a random variable and its variation combines with that of the noise background. A reasonable assumption here is that S and X_n are independent. The variance of the signal-plus-noise distribution is now the sum of the variances of its parts (Equation A.33 on page 235):

$$\mathrm{var}(X_s) = \mathrm{var}(X_n) + \mathrm{var}(S) \geq \mathrm{var}(X_n).$$

Any variation in the signal or in its detection makes $\mathrm{var}(X_s) > \mathrm{var}(X_n)$. Equality of variance holds only when the signal is a fixed nonrandom quantity.

Another cause of differences in signal and noise variance is a direct consequence of the mechanism by which observations of X_n or X_s are generated. The mean and variance of many random processes are related, so that random variables with larger means also have larger variances. If the response to the stimulus is generated by such a process, then the larger mean of X_s also gives it a greater variance. As a specific example, suppose that X is observed by counting the number of discrete events (perhaps neural responses) during a fixed interval of time. The rate at which these events occur is larger when the signal is present than when it is not, so the count is, on average, bigger when the signal is present. *Counting processes* of this type have been widely studied. For many of them, the variance of the number of events is approximately proportional to the mean.[2] Although counting processes do not produce true Gaussian distributions, they are usually closely approximated by Gaussian distributions whenever the number of counts is large. Because of the relationship between the mean and the variance, the unequal-variance model must be used.

In other situations one might expect the noise variance to exceed that of the signal. Sometimes the signal event acts to reduce the diversity of an original distribution. Consider a word recognition experiment in which the subject identifies words that were presented in the first part of the experiment, and suppose that the subject does this by estimating how recently he or she has heard the target word. The old, or signal, words have all been presented recently and so have a relatively tight distribution of ages, but the new, or noise, words have a great range of ages. Some words have been seen or heard only a few hours ago, while other words are days, months, or years old. When the words are studied, values from this highly variable distribution are replaced by much more similar values that refer to the experimental presentation.

The unequal-variance Gaussian model depends on three parameters: the two distributional parameters, μ_s and σ_s^2, and the criterion λ. Thus, its parameters cannot be determined by a single yes/no detection study, which yields but two independent results (f and h). Fitting the unequal-variance model requires either several conditions varying in bias or the rating-scale experiment of Chapter 5.

The theoretical operating characteristic for the unequal-variance model is constructed by converting the response rates to Gaussian coordinates as in Section 3.2. As in the equal-variance model, the false-alarm rate depends directly on the criterion:

$$P_F = \Phi(-\lambda) \qquad \text{and} \qquad z_F = -\lambda.$$

The signal distribution now involves its variance. Using the rule for finding area under a normal distribution with nonunit variance (Equation A.46 on page 240),

$$P_H = P(X_s > \lambda) = 1 - \Phi\left(\frac{\lambda - \mu_s}{\sigma_s}\right) = \Phi\left(\frac{\mu_s - \lambda}{\sigma_s}\right).$$

---

[2]The simplest of the counting processes is the *Poisson process*, in which the counted events are postulated to occur independently at a constant rate. It gives rise (not surprisingly) to Poisson distributions for X_n and X_s. The mean and variance of a Poisson random variable are equal.

Inverting this relationship gives the ordinate of the operating characteristic:

$$z_H = Z(P_H) = \frac{\mu_s - \lambda}{\sigma_s}.$$

Finally, substituting $-z_F$ for λ gives the equation of the isosensitivity function:

$$z_H = \frac{1}{\sigma_s} z_F + \frac{\mu_s}{\sigma_s}. \tag{3.4}$$

This function is linear, with a slope that is the reciprocal of the signal standard deviation. In the case depicted in Figure 3.8, where $\sigma_s = 1.5\sigma_n$, the slope is $1/1.5 = 2/3$.

When the isosensitivity function for a model with unequal variance is translated from Gaussian coordinates back to probability coordinates, the resulting operating characteristic is not symmetric about the minor diagonal of the unit square (the line from lower right to upper left), as it was in the equal-variance case. If $\sigma_s > 1$, then the function has a form like that shown in Figure 3.9. The line rises sharply from (0, 0) as the criterion drops through the signal distribution without reaching much of the noise distribution, then turns more slowly toward the right as the noise distribution is passed while there is still appreciable area under the signal distribution. Eventually the curve approaches (1, 1) when the criterion is below both distributions.

The operating characteristic in Figure 3.9 has several disconcerting features. The most obvious of these is that it dips below the diagonal at the upper right. This dip suggests that for certain criterion positions the false-alarm rate exceeds the hit rate. More subtle is the fact that the slope of the operating characteristic goes from shallower to steeper in this region, a condition that is necessary for the dip to occur. Operating characteristics in which the slope changes nonmonotonically like this are said to be *improper*. Those with curves that progress from (0, 0) to (1, 1) with ever-decreasing slope are said to be *proper*. An improper operating characteristic is a sign that the observer is using a response rule that does not make optimal use of the available information. In the case of the unequal-variance Gaussian model, the improper operating characteristic indicates that a simple criterion applied to the axis is not the best way to make the response. A superior response rule will be described in Section 9.3.

On first being introduced to the unequal-variance Gaussian model, one is inclined to worry about the dip below the diagonal and the apparently perverse behavior that it implies. However, to fret much about it is to take the model too seriously. The model is an useful description of detection behavior, but cannot be taken as mathematical truth. There are many

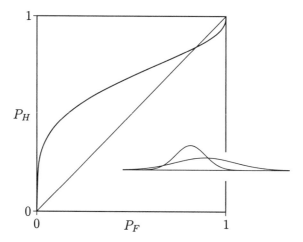

Figure 3.9: *The isosensitivity function for the unequal-variance Gaussian model with $X_n \sim \mathcal{N}(0,1)$ and $X_s \sim \mathcal{N}(1,4)$. The inset shows the distributions of X_n and X_s.*

circumstances in which the nonunit slope of a transformed operating characteristic implies that an unequal-variance representation is appropriate, but where no evidence of the dip is present. In fact, to observe the dip in real data is almost impossible. It would occur only for conditions where the observer was so biased toward YES responses that both the hit rate and the false-alarm rate were almost, but not quite, equal to one. An accurate determination of such an extreme point would require an enormous sample of data. It is doubtful that an observer could be induced to perform in a stable way in such extreme conditions for the requisite number of observations.

3.5   Fitting an empirical operating characteristic

Fitting the unequal-variance signal-detection model requires a set of conditions that differ in bias but not in the detectability of the signal. These conditions can be created by keeping the signal and noise events the same while inducing the observer to alter the criterion. There are several ways to produce this shift. Practiced observers can respond to instructions to change their detection performance in a way that increases or decreases the hit rate or false-alarm rate. Less trained observers will change the criterion in response to the proportion of signals. The analysis of the ideal observer in Section 2.5 indicated that bias should be shifted with the signal frequency to maximize correct responses (Equation 2.15). Real observers shift their performance in this direction, although usually not to the optimal extent. When signals are rare, fewer YES responses are made, lowering both the hit rate and the false-alarm rate. When signals are common, the bias is shifted to increases the number of YES responses. Thus, running sessions with different proportions of signals will give different points on the operating characteristic. Bias can also be manipulated by assigning payoffs to the various types of responses, as in the example of Section 1.2. Each of these manipulations concerns the observer's response behavior only and can typically be made without substantially altering the sensitivity.

The first thing to do with such data is to look them over for anomalies or inconsistencies that could indicate something wrong with the study. Plot the hit rate against the false-alarm rate and verify that the general form of an operating characteristic is obtained. Problems are indicated by points for which the false-alarm rate exceeds the hit rate or by a failure of the order of the conditions on the operating characteristic to match the manipulation used to shift the bias. The sampling fluctuation associated with any set of empirical observations can lead to some reversals or to points where $h < f$, particularly when the amount of data is small, but large discrepancies are unlikely. When they happen, it is possible that the observer misunderstood the task or was behaving perversely, but that is unlikely in a well-run study. A more common cause of these irregularities is a mistake by the researcher in recording the data or in entering them into a computer program for analysis. Such errors should be ruled out before taking an irregular point too seriously.

When the data have been deemed satisfactory, turn to the analysis. It is helpful to organize the analysis in five steps:

1. Convert the data from hit and false-alarm rates h and f to the transformed values $Z(h)$ and $Z(f)$, using the table of the Gaussian distribution on page 250 or an equivalent computer program. Plot these points as an operating characteristic in Gaussian coordinates. Be accurate—it helps to use a full sheet of graph paper with closely spaced grid lines.

2. Evaluate the Gaussian model. If the points fall in a straight line (except for what can be deemed sampling error), then the use of this model is justified. If they curve, then consider a model based on another distribution. Under most circumstances a visual evaluation is sufficient (formal statistical tests are covered in Section 11.5). Usually, the Gaussian model will be adequate.

3. Fit a straight line to the set of points. Be sure to try the line at 45° that corresponds to the equal-variance model. When there is little

scatter in the data, the line can be drawn by eye without serious error. When the data are more scattered, some form of statistical fitting procedure is better, although rough estimates still can be made by eye. A transparent 45° drafting triangle is very useful here. Decide whether the 45° is adequate (some statistical procedures are discussed in Section 11.6). Otherwise conclude that the variances are unequal and that a line of some other slope is needed.

4. Estimate the parameters of the signal distribution from the fitted line. If the chosen line has a slope of one, then estimate d' by the intercept of the function, as described in Section 3.2. If the line does not lie at 45°, then find the slope b and intercept a of the equation

$$Z(h) = bZ(f) + a.$$

Match the slope and intercept to Equation 3.4 to estimate the Gaussian model's parameters:

$$\widehat{\mu}_s = a/b \quad \text{and} \quad \widehat{\sigma}_s = 1/b. \tag{3.5}$$

When working from a graph on which the line has been drawn, it is easier to forget about the slope and intercept and instead note the points x_0 and y_0 where the line crosses the horizontal and vertical axes, respectively. The parameters of the line are

$$a = y_0 \quad \text{and} \quad b = -y_0/x_0, \tag{3.6}$$

and those of the detection model are

$$\widehat{\mu}_s = -x_0 \quad \text{and} \quad \widehat{\sigma}_s = -x_0/y_0. \tag{3.7}$$

5. To estimate the criterion λ for a condition, the observed point must be translated to one on the fitted line. A full analysis here takes account of the accuracy with which each coordinate is observed. However, for most purposes it is sufficient to take the bias from the point on the operating characteristic that is closest to the observation. Frequently this can be done by sketching a line perpendicular to the line that goes through the point. Details and formulae are given in Section 4.5. Once the point is chosen, $\widehat{\lambda}$ is minus the abscissa of the point.

Example 3.2: Three carefully measured detection conditions give the pairs of false-alarm and hit rates (0.12, 0.43), (0.30, 0.76), and (0.43, 0.89). Does a Gaussian model fit these data? If it does, then estimate its parameters.

Solution: First sketch the data in probability coordinates (not shown) and look for irregularities. Here there are none. Then transform the probabilities to Gaussian coordinates:

f	h	$Z(f)$	$Z(h)$
0.12	0.43	-1.17	-0.18
0.30	0.76	-0.52	0.71
0.43	0.89	-0.18	1.23

The transformed points are plotted in Figure 3.10. A line at $45°$ (dashed in the figure) is clearly unsatisfactory, so the equal-variance Gaussian model does not apply. However, they lie so close to another straight line that it can be drawn through them by eye (the solid line in the figure). Evidently, the unequal-variance Gaussian model fits these data. Because the slope is greater than one, the signal distribution has a smaller variance than the noise distribution. Reading the graph carefully, the line crosses the horizontal axis at $x_0 = -1.04$ and the vertical axis at $y_0 = 1.46$. The slope and intercept of the line are found from these crossing points (Equations 3.6):

$$a = y_0 = 1.46 \qquad \text{and} \qquad b = \frac{y_0}{-x_0} = \frac{1.46}{1.04} = 1.40.$$

These values can be converted to estimates of the parameters of the signal distribution using Equations 3.5, or they can be found directly from the crossing points (Equations 3.7):

$$\widehat{\mu}_s = -x_0 = 1.04 \qquad \text{and} \qquad \widehat{\sigma}_s = \frac{-x_0}{y_0} = \frac{1.04}{1.46} = 0.71.$$

Finding the criteria associated with the conditions is simple here. The line fits so accurately that its closest approaches to the three points are at the abscissas already calculated as $Z(f)$. Accordingly, the criteria are 1.17, 0.52, and 0.18.

The problem of finding the best-fitting line deserves comment. The fact that one is fitting a straight line to a group of points makes it tempting to use the simple fitting equations of ordinary linear regression. One would write z_H as a linear function of z_F and calculate the slope and intercept from the usual regression equations. Unfortunately, this procedure gives the wrong line and systematically biased estimates of the parameters. The difficulty lies in the way that the regression line is defined. The model for linear regression treats one variable x as a predictor and the other variable y as an outcome to be predicted from x. The outcome variable y is represented theoretically as a linear function of the predictor plus a random error e:

$$y = \alpha + \beta x + e.$$

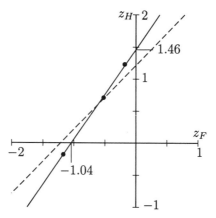

Figure 3.10: A three-point operating characteristic for Example 3.2 plotted in Gaussian coordinates. The dashed line at 45° is unsatisfactory, but the solid line from the unequal-variance model fits well.

In this regression model, x is an exact number, and all the sampling uncertainty is attributed to y. This model is a unsatisfactory representation of detection data, for which both $Z(h)$ and $Z(f)$ are subject to sampling error.

The use of the regression model here would not create problems if it did not bias the estimates of the detection parameters. The problem is a phenomenon known as *regression to the mean*. Geometrically, variability in the data acts to flatten the regression line, so that the best prediction of y is nearer (in standard deviation units) to the mean of that variable than the predictor x is to its mean. The greater the variability of the data, the more the line is flattened. Shrinking the predictions toward the mean is appropriate for the asymmetric measurement structure of regression analysis, but is incorrect for detection data. The line regressing $Z(h)$ on $Z(f)$ has a smaller slope than does a line that treats both $Z(h)$ and $Z(f)$ as subject to sampling error. Unless the data are almost error free, a regression analysis will give too small a value for b and consequently overestimate $\hat{\sigma}_s$. When there is considerable scatter in the data, this bias can lead the equal-variance model to be rejected when in fact it is appropriate.

3.6   Computer programs

The amount of calculation involved in fitting the signal-detection theory model makes it a good candidate for computerized calculation. The calculations for a single detection condition are easy to implement. Algorithms for the functions $Z(f)$ and $Z(h)$ are well established, and their values are directly available in some higher level languages. Using them, Equations 2.3 and 2.4 can be calculated directly.

When three or more conditions with different bias are to be fitted, the task is considerably harder. It is no longer possible to convert the proportions directly into parameter estimates, but a program must, in effect, find the operating characteristic that fits the data best, even though it may not pass exactly through any of the observed points. However, because the probabilistic structure of the signal-detection model is well defined, it can be fitted with the standard statistical technique known as maximum-likelihood estimation. Several programs that make these calculations have been published (see reference notes). These usually report most or all of the statistics that will be discussed in Chapter 4. One of the great advantages of the maximum-likelihood procedure is that it gives values for the standard errors of the parameter estimates. The use of these estimates will be discussed in Chapter 11.

When using one of these programs, it helps to know a little about what they are doing. There are no equations that directly give estimates of the parameters of the Gaussian model in terms of the data (e.g., as there are in multiple regression). Instead, the signal-detection model is fitted by an iterative algorithm. The programs start by making some guess at the parameter values—not necessarily a very good one. Then it adjusts these values to improve the fit. This process of adjustment, or iteration, is repeated (generally out of sight of the user) until the estimates cease to improve, at which time the results are reported and the values of any summary measures are calculated. Many programs report the number of iterations required to complete the process or the criterion of change used to decide when to stop.

For most sets of data, this procedure runs successfully and delivers good estimates. However, with certain very irregular sets of data, the algorithm can break down and fail to find a solution—a failure to converge, as it is called. Exactly how the program handles this contingency depends on the implementation, and some versions of the algorithm are more robust than others. Failure of the estimates to converge is uncommon with signal-detection data. It most often occurs when the observations are very much inconsistent with the signal-detection models, for example, with points lying very far from a line in Gaussian space. Thus, when a program fails to converge, it is advisable to review the data and see if they have been entered incorrectly or if they are sufficiently at odds with the signal-detection model that any parameters obtained by fitting that model will be meaningless. If it is necessary find estimates for such data nonetheless, it may be possible to revise the starting point for the search to one from which it will converge.

Fitting a line to the data by eye may give a good place to start. It is also worth trying another program, as minor differences in the way the program selects its starting point or implements the iterative algorithm give them different sensitivities.

Reference notes

Some approximations for calculating $\Phi(z)$ and $Z(p)$, sufficiently accurate for calculation of d' and the like, are given by Zelen and Severo (1964). The standard maximum-likelihood estimation algorithm used to fit the signal-detection model to several conditions was originally published by Dorfman and Alf (1968a, 1968b). The background to these methods are found are discussed in most advanced statistics texts. Several implementations of the procedure have been published, and a summary of programs is given in Swets (1996). I will make my version of these programs available on the web site mentioned in the preface.

Exercises

3.1. Suppose that $\mu_s = \sigma_s = 1.5$ in the Gaussian signal-detection model.

a. Draw the isosensitivity function in Gaussian coordinates.
b. Convert this function to an operating characteristic in probability coordinates.

3.2. Two detection conditions give estimates of $\widehat{d'_1} = 1.05$, $\widehat{\lambda}_1 = 1.03$ and $\widehat{d'_2} = 1.70$, $\widehat{\lambda}_2 = 0.38$. Use these results to fit an unequal-variance model to the pair of conditions. Draw the operating characteristic.

3.3. Six detection conditions (A to F) are run with the same signal and observer but with the bias manipulated. The response frequencies observed are

	Noise		Signal	
	YES	NO	YES	NO
A	264	36	294	6
B	168	132	273	27
C	102	198	252	48
D	30	270	198	102
E	17	283	171	129
F	2	298	108	192

a. Estimate d' and $\log \beta$ for each condition, based on the equal-variance Gaussian model. Is there a pattern to these values? Why is this analysis unsatisfactory?

b. Sketch (by eye) an isosensitivity function for these data in standard coordinates.

c. Plot the data in Gaussian coordinates and draw a straight line by eye through them.

d. Use your line to decide whether $\sigma_s^2 = \sigma_n^2$.

e. Estimate the parameters of whichever model you chose in the previous step.

3.4. Use a program to fit the data from Problem 3.3. Compare the results to the estimates made in this problem.

3.5. Suppose that the hit rate and false-alarm rate for detection of visual stimulus are collected from five different subjects. Explain why it would be inappropriate to use the five points to create an operating characteristic as described in this chapter.

Chapter 4

Measures of detection performance

In the equal-variance Gaussian model, the distributions of X_n and X_s differ only by the parameter d'. This quantity easily summarizes detection performance. For the unequal-variance model, where the distributions differ in both mean and variance, finding a single measure of detectability is less obvious. The differences between the noise and signal configurations now involve two parameters, μ_s and σ_s^2. These cannot be summarized by one number. Any choice of a single detection index necessarily emphasizes some aspects of the difference and minimizes others. This chapter describes these alternatives. The first three sections present four measures of detection. No one of these measures clearly dominates the others in all situations. Each has its own advantages and liabilities, which are summarized, with recommendations, in Section 4.4. Although most studies are concerned primarily with measuring detectability, questions concerning bias also arise. Section 4.5 discusses the possible measures. Finally, although the prototypical signal-detection study is one in which extensive data are collected from a single observer, many studies are run in which smaller amounts of data are collected from many observers. The combination of results from different observers presents special problems that are treated in Section 4.6.

The issues discussed in this chapter apply to any study in which sufficient data are available to draw an operating characteristic. These data could come from manipulations of the bias in separate conditions or from the rating-scale procedure that will be discussed in Chapter 5. As an example for use in this chapter, Figure 4.1 gives the results of four conditions with responses from 200 trials in each condition. The figure also shows the operating characteristic in standard and Gaussian coordinates. A look at the transformed operating characteristic shows that the Gaussian model

60

Set	F.A.	Hits	f	h	$Z(f)$	$Z(h)$
1	33	81	0.165	0.405	−0.974	−0.240
2	55	121	0.275	0.605	−0.598	0.266
3	86	127	0.430	0.635	−0.176	0.345
4	115	150	0.575	0.750	0.189	0.674

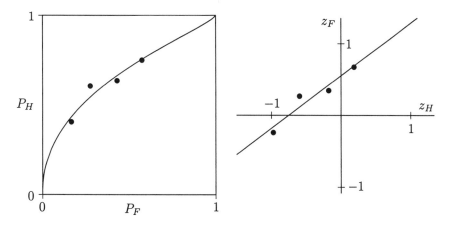

Figure 4.1: Detection performance in four bias conditions, each with 200 signal trials and 200 noise trials. The plots show the operating characteristic fitted from the unequal-variance Gaussian model. The fitted line, $z_H = 0.550 + 0.767z_F$, crosses the axes at $x_0 = -0.747$ and $y_0 = 0.550$.

fits adequately (the points lie roughly on a straight line) and that the equal variance model is unsatisfactory (that line is not at 45°). The slope is less than 45°, which implies that $\sigma_s^2 < \sigma_n^2$. The issue now is how to find a number that measures the detectability for these data.

4.1   The distance between distributions

Any measure of detectability that integrates across conditions must be based, at least implicitly, on a theoretical representation that ties the observations together. For data such as those in Figure 4.1, the Gaussian model is a good choice. When this model has been fitted, a natural candidate for a measure of the difference between the two distributions of this model is the difference between their means, often denoted by[1]

$$\Delta m = \mu_s - \mu_n. \tag{4.1}$$

[1]The uppercase Greek delta here is a difference operator, and so Δm is not the product of Δ and m, but a difference between two values of the mean m (or μ).

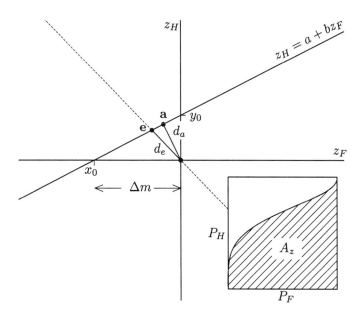

Figure 4.2: *The detection measures shown as distances Δm, d_e, and d_a on the operating characteristic in Gaussian coordinates. The inset operating characteristic shows the area A_z under the curve in probability coordinates.*

With the noise mean μ_n set to zero, Δm is the same thing as the signal mean μ_s, which, as the notation suggests, is interpreted as a difference between means. This measure is simple and is easily determined from a straight line fitted to a series of detection conditions in Gaussian coordinates. Figure 4.2 shows this measure (and other measures to be discussed later) for an unequal-variance operating characteristic, both in Gaussian and probability coordinates. The difference Δm equals that between the origin and the point x_0 on the horizontal axis where the operating characteristic crosses it (recall Equation 3.7 on page 54).

Equation 4.1 is a theoretical statement about the parameters of the model. A comparable equation expresses the relationship among the estimates of these quantities:

$$\widehat{\Delta m} = \widehat{\mu}_s - \widehat{\mu}_n.$$

Because μ_n is by definition zero, the relationship becomes simply $\widehat{\Delta m} = \widehat{\mu}_s$. The value of $\widehat{\Delta m}$ is taken directly from the calculations in Section 3.5.

Although it is important not to confuse the theoretical quantities with their estimates, the equations that relate the theoretical quantities in this chapter are identical to their counterparts for the estimates. Little is gained by writing them twice. Hence, to avoid duplication and notational clutter, the distinction between a quantity and its estimate is not preserved notationally here. The context in which an equation is used makes it clear whether the theoretical parameters of the signal-detection model or their estimates from data are under consideration. In applications—say, when discussing data—the estimates are often not specifically marked, but in theoretical discussions it is valuable to preserve the distinction.

Example 4.1: Find Δm from the four conditions in Figure 4.1.

Solution: Figure 4.1 shows the frequencies converted to the proportions f and h and transformed into $Z(f)$ and $Z(h)$. The straight line $z_H = a + bz_F$ fitted to the transformed points (using a computer program) is

$$z_H = 0.550 + 0.736z_F,$$

with axis crossing points at $x_0 = -0.747$ and $y_0 = 0.550$. The operating characteristic line crosses the z_F axis at $-\Delta m$. This value can be read directly from the graph or it can be calculated from the numerical slope and intercept:

$$\Delta m = -x_0 = \frac{a}{b} = \frac{0.550}{0.736} = 0.747.$$

A convenient feature of Δm as a detection measure is its closeness to the signal-detection model. It is directly equivalent to a parameter of the Gaussian model, so that it is easy to interpret once that model is understood. When the variances of the signal distribution and the noise distribution are the same, Δm has the same value as d', so it is interpreted in the same way, particularly when the difference between the variances is small.

As a measure of detection, Δm suffers from two disadvantages, one related to its interpretation, the other statistical. The interpretational difficulty occurs because Δm does not completely capture the extent to which the distributions overlap. The magnitude of the overlap depends on both the difference between means and the variance of the signal distribution. Figure 4.3 illustrates the problem. In both panels, $\Delta m = 2$. In the top panel σ_s is half the size of σ_n. There is not a great deal of overlap in the distributions and a good observer can be correct as much as 92% of the time. In the bottom panel, the relative sizes of the standard deviations are reversed, so σ_s is twice σ_n. There is more overlap and the observer now can at best be correct only 77% of the time. In spite of the equal value of Δm, it seems unreasonable to treat the two signals as equally detectable.

The statistical difficulty with Δm is of greatest concern when there is substantial variability in the data. The extent to which a quantity varies accidentally from one replication to another under identical conditions is

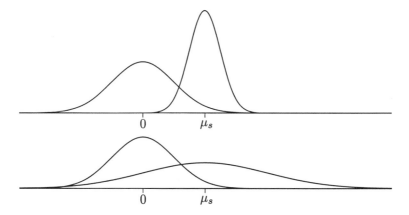

Figure 4.3: Two pairs of noise and signal distributions with $\sigma_n^2 \neq \sigma_s^2$. Although $\Delta m = 2$ for both pairs, the overlap of the distributions is greater and optimal performance is lower in the pair for which σ_s^2 is large.

measured by its standard error. The standard error of Δm is larger than it is for several comparable measures. The techniques in the chapter on statistical treatment can be used to make exact calculations (see Section 11.3), but the problem can be understood geometrically. A line fitted to a swarm of points is most accurately determined in the middle of the swarm. Its accuracy is greatest at the mean of the observations and falls off as one moves away from this center, an effect discussed for regression lines in many statistics books. A measure that depends on the position of a single point on the line can be determined with the least error when that point lies in the middle of the points used to estimate the line. However, most of the data points in a typical experiment fall to the right of the intercept that determines Δm (e.g., as in Figure 3.7 on page 48). The place where the isosensitivity line crosses the horizontal axis is away from the center of the data, making Δm less accurate as a measure of the line's location than a point in the middle of the data.

4.2   Distances to the isosensitivity line

As Figure 4.2 shows, the detection measure Δm is the horizontal distance between the origin in Gaussian coordinates and the isosensitivity line. The detectability of the signal, in this sense, is measured by how far this line is from the origin. The point on the line to which this distance is measured corresponds to a condition with a hit rate of $P_H = 1/2$, which is a peripheral condition in most studies. One way to get a better measure is to derive it from a more central point, which will be more stable. Two measures have been proposed that are based on this idea.

One measure, known as d_e, derives from the point corresponding to a condition with symmetrical error rates. For this balanced conditions, the false-alarm rate equals the miss rate. It is likely to fall in the middle of the range of points produced by bias manipulations in the typical experiment. At this point, the equality of the error rates means that $z_H = -Z[P(\text{miss})] = -z_F$. The point, labeled e in Figure 4.2, lies at the intersection of the operating characteristic $z_H = a + bz_F$ and the minor diagonal $z_H = -z_F$. Solving these simultaneous equations shows that they intersect at $e = (-z_e, z_e)$, with $z_e = a/(1 + b)$. One could express the relationship of this point to the origin 0 either by the distance from 0 to e or by z_e directly. However, it is useful to make the measure coincide with d' when the equal-variance model holds. For that model, $a = d'$ and $b = 1$, so that $z_e = d'/2$. Accordingly, define the measure d_e to be twice z_e. This measure can be written in terms of the slope and intercept, the parameters of the signal-detection model, or the axis crossing points of the operating characteristic (using Equations 3.4 and 3.6):

$$d_e = \frac{2a}{1+b} = \frac{2\mu_s}{1+\sigma_s} = \frac{2x_0 y_0}{x_0 - y_0}. \tag{4.2}$$

When $\sigma_s = 1$, then $d_e = d'$.

The second measure, known as d_a, derives from the point a on the operating characteristic that is closest to the origin in Gaussian coordinates. The point of closest approach of a line to a point lies in a direction perpendicular to the line (see Figure 4.2). Lines perpendicular to the isosensitivity line $z_H = a + bz_F$ have a slope of $-1/b$, so the closest approach to the origin is along the line $z_H = -z_F/b$, which has this slope and passes through the origin. Solving this pair of simultaneous equations, as in finding e, shows that the lines intersect at the point $a = [-ab/(1 + b^2), a/(1 + b^2)]$. The distance from the origin to a is $a/\sqrt{1 + b^2}$. Again, it is helpful to bring the measure in line with d', which is done by multiplying the distance by $\sqrt{2}$. The closest-approach measure is

$$d_a = \frac{\sqrt{2}\,a}{\sqrt{1+b^2}} = \frac{\sqrt{2}\,\mu_s}{\sqrt{1+\sigma_s^2}} = \frac{\sqrt{2}\,x_0 y_0}{\sqrt{x_0^2 + y_0^2}}. \tag{4.3}$$

Example 4.2: Calculate the distance detection statistics for the data in Figure 4.1.

Solution: Working from the slope and intercept found in Example 4.1, and using Equation 4.2, the distance measure d_e is

$$d_e = \frac{2a}{1+b} = \frac{2 \times 0.550}{1 + 0.736} = 0.634.$$

Likewise, from Equation 4.3, the distance measure d_a is

$$d_a = \frac{\sqrt{2}\,a}{\sqrt{1+b^2}} = \frac{1.414 \times 0.550}{\sqrt{1+(0.736)^2}} = 0.626.$$

Although d_a is perforce a little smaller than d_e (it is based on the closest approach to the origin), there is little difference between them. They both differ from the estimate $\Delta m = 0.747$ found for these data in Example 4.1.

Starting with a sketched line, one can get the same answers from the crossing points. For example,

$$d_a = \frac{\sqrt{2}\,x_0 y_0}{\sqrt{x_0^2 + y_0^2}} = \frac{1.414 \times (-0.747) \times 0.550}{\sqrt{(0.550)^2 + (-0.747)^2}} = 0.626.$$

Although measures d_a and d_e are not as tightly linked to the parameters of the Gaussian signal-detection model as is Δm, they make better indices of detection. There is not a great deal of difference between them. The symmetrical error structure associated with **e** makes d_e a plausible choice. The point **a** does not have such a simple interpretation, and when thought of as describing the outcome of a study, d_a seems a little arbitrary. However, it turns out that d_a is closely related to the measure A_z that is discussed in the next section, which gives it an advantage.

In most studies, the points **e** and **a** that determine either d_e and d_a fall near the middle of the data points that are used to determine the operating characteristic. These points are more stable and less influenced by accidents than points farther away from the center. Thus, except with very unusual sets of data, either d_e or d_a is statistically superior to Δm.

4.3   The area under the operating characteristic

The three detection measures discussed so far all derive from the Gaussian model: Δm is a parameter of the model and d_s and d_a depend on the fact that the isosensitivity line in Gaussian coordinates is straight. The final detection measure considered here is less directly tied to that model, at least superficially. It is simply the area under the isosensitivity curve when it is plotted in ordinary probability coordinates. This area, denoted generically by A, is shaded in the inset to Figure 4.2.

A look at the sequence of isosensitivity contours for the equal-variance model plotted in Figure 3.4 on page 43 shows how as d' increases, the operating characteristics shift up and to the left. The area below the chance function ($d' = 0$) is $\frac{1}{2}$, and as d' gets large it increases to 1. These characteristics apply to models other than the equal-variance Gaussian: the area is $\frac{1}{2}$ when the responses are unrelated to the stimulus, it increases with the detectability of the signal, and it asymptotes to unity for very strong signals.

The value of A has an interpretation in procedural terms that is one of its most attractive features. It is equal to the probability of a correct response in the *two-alternative forced-choice experiment* that will be discussed in Chapter 6. In that procedure, the observer is presented on each trial with two stimuli, one that contains the signal and one that does not. The observer's task is to decide which is which. As will be shown in Section 6.4, a result known as the *area theorem* shows that the probability of a correct response by an observer who is performing optimally in the forced-choice task is equal to the area under that observer's isosensitivity curve in a detection experiment using the same stimuli. This interpretation of A as a correct-response probability is easy to understand.

As just defined, the area measure A does not depend on any particular assumptions about the form of X_s and X_n, or even on the use of a distribution-based model. If an operating characteristic can be determined, then the area beneath it can be calculated without reference to a particular detection model. In this sense, A is sometimes described as a "nonparametric" or "distribution-free" measure of detection. As will be seen, this designation must be qualified.

From a mathematical point of view, the area under the theoretical isosensitivity curve is calculated by an integral. Suppose that for a particular detection model, the operating characteristic is given by the function $P_H = R(P_F)$. By definition, the area A under this line is the integral of $R(p)$ between $p = 0$ and $p = 1$:

$$A = \int_0^1 R(p)\, dp. \tag{4.4}$$

For any model that is specific enough to determine the function $R(p)$, this area can be calculated, either analytically or numerically. As in the formal analysis of Section 3.2, write the area in terms of the functions that relate the criterion to the false-alarm rate and hit rate, $P_F = F(\lambda)$ and $P_H = H(\lambda)$ (Equation 3.2 on page 45). The equation of the operating characteristic is $R(p) = H[F^{-1}(p)]$, and the area is

$$A = \int_0^1 H[F^{-1}(p)]\, dp. \tag{4.5}$$

This integral gives the area.

Now turn to the Gaussian model, for which the hit and false-alarm functions are

$$F(\lambda) = \Phi(-\lambda) \qquad \text{and} \qquad H(\lambda) = \Phi\left(\frac{\mu_s - \lambda}{\sigma_s}\right).$$

A calculation that will be presented in Section 6.4 shows that, with these functions, the area under the operating characteristic is

$$A_z = \Phi\left(\frac{\mu_s}{\sqrt{1 + \sigma_s^2}}\right). \tag{4.6}$$

The subscript z here is a reminder that this area is derived from the Gaussian model. In terms of the isosensitivity function and the crossing points,

$$A_z = \Phi\left(\frac{a}{\sqrt{1 + b^2}}\right) = \Phi\left(\frac{x_0 y_0}{\sqrt{x_0^2 + y_0^2}}\right). \tag{4.7}$$

Example 4.3: Find the area under the Gaussian isosensitivity contour for the data in Figure 4.1.

Solution: Using the estimated slope and intercept obtained in Example 4.1, Equation 4.7 gives

$$A_z = \Phi\left(\frac{a}{\sqrt{1 + b^2}}\right) = \Phi\left(\frac{0.550}{\sqrt{1 + (0.736)^2}}\right) = \Phi(0.443) = 0.671.$$

The term within the distribution function Φ in Equation 4.6 is the same as the distance from the origin to the point **a** where the operating characteristic passes closest to the origin (Equation 4.3). As a result, the area measure is related to the distance measure d_a by

$$A_z = \Phi(d_a/\sqrt{2}). \tag{4.8}$$

The area measure lies between $1/2$ and 1 and the distance measure between 0 and ∞. The relationship between them looks like half of a cumulative Gaussian distribution function (Figure 4.4). When the equal-variance model holds, the distance-based measures are identical, and d' can be used instead of d_a in Equation 4.8 or on the abscissa of Figure 4.4.

The fact that area A is defined very generally for any operating characteristic suggests that it might be possible to calculate this area directly from a set of observed data without going through a theoretical model. It is possible to do something of the sort, although the resulting area has some serious defects. The idea is to connect the observed points by straight lines, then break up the area below them by trapezoids whose areas are easily

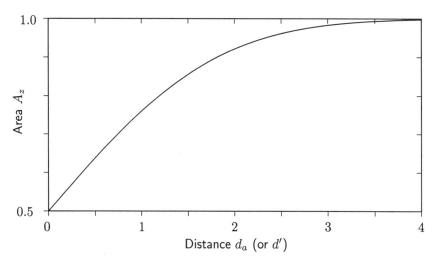

Figure 4.4: The relationship between the area A_z under the Gaussian operating characteristic and the distance d_a (or d' for the equal-variance model).

calculated. The procedure is illustrated by a two-condition study that is analyzed in Figure 4.5. The two observations lie at $f_1 = 0.133$ and $h_1 = 0.333$ and at $f_2 = 0.333$ and $h_2 = 0.660$. The empirical operating characteristic goes from the origin to the first point, then to the second point, and finally to the point $(1, 1)$. The area under this curve is the sum of the area of the triangle A_1 and that of the two trapezoids A_2 and A_3. The area of a triangle is half the product of its base and its height, while the area of a trapezoid is the base times the average of the two heights. The three sections sum to give the total shaded area in Figure 4.5:

$$A_{trap} = A_1 + A_2 + A_3 = \frac{f_1 h_1}{2} + \frac{h_1 + h_2}{2}(f_2 - f_1) + \frac{h_2 + 1}{2}(1 - f_2).$$

Substituting the numerical values gives the area $A_{trap} = 0.649$. For a study with n points, A_{trap} is given by a generalization and rearrangement of the two-point formula:

$$A_{trap} = \frac{1}{2}\left[1 + \sum_{i=2}^{n}(h_{i-1}f_i - h_i f_{i-1}) + h_n - f_n\right]. \tag{4.9}$$

Although A_{trap} is easier to calculate than the other measures, its usefulness is limited by two problems. First, in many sets of data the points do not form a monotonically increasing sequence. Without such a sequence, the trapezoids cannot be cleanly drawn. For example, the four points in

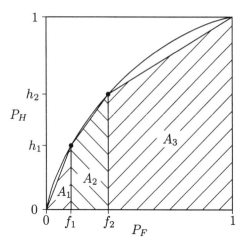

Figure 4.5: *The trapezoidal area A_{trap} under an empirical isosensitivity curve. The smooth curve gives the equal-variance Gaussian function fitted to the same data.*

Figure 4.1 do not form a proper operating characteristic, but one whose slope goes from steep to flatter to steep again. Still worse are the data in Example 3.1 on page 47. A line connecting the points in Figure 3.7 is not monotonic and does not correspond to a reasonable operating characteristic. In both cases, some sort of smoothing of the data is necessary. This smoothing cannot be done without an idea of what the operating characteristic should look like. The best way to regularize the results is to fit them with a theoretical model. In Figure 4.5, this smoothing has been done by fitting the equal-variance Gaussian model to the two points, and the result is shown by the smooth line. The area under this line is $A_z = 0.685$, which is greater than $A_{\text{trap}} = 0.649$. Other methods of smoothing could be used, but each implies something about what the true operating characteristic should look like. There is no good reason to prefer any of these over the Gaussian fit.

The second problem with A_{trap} is its bias. The empirical operating characteristic was formed by connecting the points by straight lines. However, regular isosensitivity contours are almost always bowed and concave downward, like those of the Gaussian model. Even when straight line operating characteristics are found (as they are for the finite state models of Chapter 8), the transitions from one segment to another are unlikely to be at the observed data points. As a result, the function formed by connecting the data points lies below the true operating characteristic—it is actually the lower bound of all possible regular operating characteristics that pass through the points. The area A_{trap} is an underestimate of the true area.

The bias in A_{trap} would be less of a matter for concern were its magnitude not dependent on characteristics of the experiment itself. The amount of bias is small when many points on the empirical function have been found and their false-alarm rates are spread out evenly between zero and one. The bias is larger when only a few points are measured or they are bunched together. For example, the bias of A_{trap} in Figure 4.5 would be reduced by adding a point that lies on the true function and has a false-alarm rate between f_2 and 1. This dependence of the area on the conditions used in the experiment undercuts the distribution-free claims for A_{trap}. One way to eliminate the bias is to use the same approach that was necessary to deal with the sampling irregularities. The calculation must be based on a model such as the Gaussian model. Even if this model is not quite correct, the errors that it introduces are surely less than the inevitable negative bias of A_{trap}.

When only a single pair of hit and false-alarm probabilities is observed, any estimate of the area under the operating characteristic requires a considerable extrapolation, as the entire curve must be reconstructed from a single point. Of necessity, one must go beyond the data here. A good estimate of the area, derived from the Gaussian model in Equation 4.8, is $A_z = \Phi(d'/\sqrt{2})$. Some other measures of area have been proposed that are sometimes described as better approximations to the true area than A_z. Perhaps the most compelling of these are the measures based on the triangles or trapezoids formed by passing lines from the origin **0** or the point $\mathbf{1} = (1, 1)$ through the observed point $\mathbf{p} = (f, h)$ (Figure 4.6). One such measure, often denoted A', averages the area under the line extended from **0** through **p** to the point marked **b** in the figure with the area under the line extended from **1** through **p** to **a**:

$$A' = \tfrac{1}{2}[(\text{area below } \mathbf{0} \text{ to } \mathbf{b} \text{ line}) + (\text{area below } \mathbf{a} \text{ to } \mathbf{1} \text{ line})].$$

An analysis of the geometry involved shows that

$$A' = 1 - \tfrac{1}{4}\left[\frac{1-h}{1-f} + \frac{f}{h}\right]. \tag{4.10}$$

Although the average A' does not equal the area under the operating characteristic associated with any particularly natural detection process, the two areas that are averaged do have interpretations in terms of the threshold models discussed in Chapter 8 (see Figure 8.3 on page 137).

It is sometimes asserted that because Equation 4.10 has been developed from the areas directly, it does not involve assumptions about the distributions of X_n and X_s. By this argument, A' is "nonparametric" and preferable to measures such as A_z, which are based on the Gaussian distribution. However, this argument is not correct. One cannot extrapolate

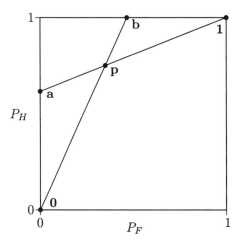

Figure 4.6: *Trapezoidal areas constructed from a single observation with $h = 0.75$ and $f = 0.35$. The letters* **0**, **a**, **b**, **p**, *and* **1** *identify points.*

from one point to an area without assumptions, and those that underlie A' are at least as stringent as those that underlie A_z (see reference note). Moreover, when data from several points are available, the operating characteristic usually looks more like the curve used to calculate A_z than like that implied by A'. Unless one has specific reasons to question it, A_z is preferable.

4.4   Recommendations

At this point a summary is in order, with some recommendation (which inevitably reflect my own preferences and biases). The goal of signal-detection theory, insofar as measuring detection is concerned, is to separate the aspects of the situation that depend on the intrinsic strength of the signal's effect on the observer from those that depend on the observer's decision to make a particular response. This separation cannot be made without some description of the decision process. The basic representation emphasized in this book is that of the Gaussian model. This description was chosen for several reasons. Among these are its historical importance and wide use, its simplicity and naturalness, its links to the statistical literature, and, most essential, the fact that it fits a great many sets of data well.

The Gaussian model is not the only possible representation. Different assumptions about the nature of the signal and the detection process lead to other models. Prominent here are the choice models mentioned briefly at the end of Section 2.1, the finite-state models to be described in Chapter 8, and the general likelihood approach of Chapter 9. Each of these representations suggests its own ways to measure the detectability of a signal. When a theoretical analysis based on one of these descriptions (or one not covered here) is adopted, the measures that flow from it should be used. A non-Gaussian model might also be chosen when it provides a better fit to a set of data, although one should be wary of sacrificing clarity of explanation for small improvements in empirical fit. Although the examination of one's data could indicate that a non-Gaussian model is required, these tests take a considerable amount of data and are not often run. On the whole, adhering to the Gaussian assumptions is acceptable.

That being said, consider the Gaussian case. When a single pair of hit and false-alarm rates is available, the natural statistic to use is d'. Either it or a transformation of it to A_z (Equation 4.8) measures detection. This much is straightforward. Either measure separates detection performance from the decision process as well as possible with such limited data.

When two or more points on the isosensitivity contour are available, a more complex representation is possible, and one must choose among several competing measures of detectability. The simple equal-variance model was extended to its more general form by adding complexity to the representation of the signal by changing X_s from $\mathcal{N}(d', 1)$ to $\mathcal{N}(\mu_s, \sigma_s^2)$, not by changing the decision component (the criterion λ or the bias $\log \beta$). Extracting a single number from this two-dimensional description of the signal invariably emphasizes one aspect of the stimulus over another. Each of the detection measures discussed in this chapter (and several others, not discussed) places this emphasis in a different way. It would be nice if psychological or perceptual theory could clearly identify which of these measures is most meaningful, but these theories are not sufficiently advanced. Indeed, it seems improbable that one representation can be the best description of the diverse situations to which signal-detection theory applies.

Lacking a strong psychological theory, one must turn to a measure that has satisfactory general properties. The strongest contenders are the measures related to the area under the operating characteristic. They integrate (both figuratively and literally) the performance over the range of potential biases. They are also somewhat less dependent on the details of the Gaussian model than are the distributional parameters, for most other representations have much the same area under their operating characteristic. Either the direct area A_z or its monotonic transformation to the distance d_a might be used—they are equivalent in that they imply the same ordering for any set of stimuli. Which of the two to prefer is less clear. For signals of moderate intensity they behave similarly. The relationship between them, shown in Figure 4.4, can be treated as a straight line out to about $d_a = 2$ without undue violence. Beyond that, the two measures differ mainly in the extent to which they separate easily detectable signals. Values of d_a map strong signals out to infinity, while values of A_z are bounded by one. It is not clear which behavior is preferable. On the one hand, d_a captures the fact that small improvements in performance at the high end may be very hard to come by and require a big difference in the physical stimulus. On the other hand, A_z is more stable and less influenced by relatively uninformative differences among conditions in which $h \approx 1$ or $f \approx 0$.

The other two measures Δm and d_e are less plausible choices. The instabilities noted at the end of Section 4.1 make the signal mean μ_s (or its alias Δm) clearly inferior to the other measures. The distance-based d_e is unobjectionable, but there is no reason to prefer it to d_a.

There remains a collection of measures that are not directly derived from an underlying detection model. These include the trapezoidal area A_{trap}, the average A' mentioned at the end of Section 4.3, and such measures as the difference between the hit rate and the false-alarm rate. Seemingly, these measures are not tied to any particular description of the detection process and give a way to get away from theoretical models. However, that is a false hope. Each such measure can be derived from a description of the detection process. To use a measure implies a willingness to either adopt that description or one equivalent to it. Typically, the structures implied by these measures are less plausible than one based on the type of graded evidence implied by the Gaussian models. It is simply not possible to develop measures of detectability without reference to what evidence was used for the decision and how that decision was made, either explicitly or implicitly. One has to have some sort of model or description process, and the Gaussian model is a good choice.

4.5   Measures of bias

Many detection studies are concerned only with measuring the detectability of the signal. The role of signal-detection theory here is to remove the contaminating effects of response bias and obtain cleaner results. However, sometimes comparisons of the bias in different conditions must be made. For this comparison, a measure of the bias is needed.

Measurement of bias is in some ways more difficult than measurement of detectability. Even with the equal-variance model, different measures with different properties were available. As discussed in Section 2.4, both λ_{center} and $\log \beta$ are plausible measures, but, as Figure 3.5 on page 44 showed, their isobias contours are quite unlike each other. It is difficult to argue on principle that one measure is better than the other. An understanding of the psychological factors that are involved in bias shifts could help, but such a theory is, if anything, farther away than an understanding of stimulus effects.

Somewhat surprisingly, the situation with the unequal-variance model is simpler than it is for the equal-variance model. A centered criterion such as λ_{center} is hard to justify when the distributions have unequal variance (where should the center be placed?). In contrast, the heights of the density functions at the criterion are clearly defined, as is their ratio. The logarithm of the likelihood ratio at the criterion (Equation 2.8) makes a clear and defensible measure of the bias:

$$\log \beta = \log \left[\frac{f_s(\lambda)}{f_n(\lambda)} \right] = \log f_s(\lambda) - \log f_n(\lambda).$$

Moreover, this measure is closely linked to the more general decision models that are discussed in Chapters 9 and 10, which gives it both a wider applicability and a plausible intuitive interpretation.

The likelihood ratio at the criterion, $\beta = f_s(\lambda)/f_n(\lambda)$, is determined by the shape of the operating characteristic. The numerator of this ratio is the rate at which probability is added to P_H as λ changes, and the denominator is the rate at which probability is added to P_F (recall that the probabilities are integrals of these functions; see Equations 3.1 on page 45). The ratio β of these rates gives the rate at which P_H changes with P_F, which is the slope of the operating characteristic.[2] As a result, one can actually use the slope of the operating characteristic as a measure of bias. It is, of course, equivalent to $\log \beta$.

The density functions for the unequal-variance Gaussian model (Equation A.42 on page 237) are

$$f_s(x) = \frac{1}{\sqrt{2\pi}\sigma_s} \exp\left[\frac{-(x - \mu_s)^2}{2\sigma_s^2} \right] \quad \text{and} \quad f_n(x) = \frac{1}{\sqrt{2\pi}} \exp\left[-\tfrac{1}{2}x^2 \right].$$

Substituting $x = \lambda$, taking the logarithms, and subtracting gives the bias:

$$\log \beta = \log f_s(\lambda) - \log f_n(\lambda)$$

$$= \left[-\log \sqrt{2\pi} - \log \sigma_s - \frac{(\lambda - \mu_s)^2}{2\sigma_s^2} \right] - \left[-\log \sqrt{2\pi} - \tfrac{1}{2}\lambda^2 \right]$$

$$= \tfrac{1}{2} \left[\lambda^2 - \frac{(\lambda - \mu_s)^2}{\sigma_s^2} \right] - \log \sigma_s. \tag{4.11}$$

---

[2]This result can be proved formally by differentiating the function $P_H = H[F^{-1}(P_F)]$ (Equation 3.2) using the chain rule.

Written using the slope and intercept of the isosensitivity line (i.e., substituting $\mu_s = a/b$ and $\sigma_s = 1/b$), this equation is

$$\log \beta = \tfrac{1}{2}[\lambda^2 - (a - b\lambda)^2] + \log b. \tag{4.12}$$

The bias of a point $(z_F,\, z_H)$ on the fitted line (i.e., one with $z_F = -\lambda$ and $z_H = a + bz_F$) is found from this equation to be

$$\log \beta = \tfrac{1}{2}(z_F^2 - z_H^2) + \log b. \tag{4.13}$$

Each of these equations reduces to the equal-variance form in Section 2.4 when $\sigma_s^2 = b = 1$.

To apply Equation 4.13, the point $(z_F,\, z_H)$ must be on the fitted isosensitivity line, and in Equations 4.11 and 4.12 the value of λ is that of this point. When the data are scattered around the best-fitting line, as in Figure 3.7 on page 48, observed values of $Z(f)$ and $Z(h)$ cannot be substituted for z_F and z_H, nor can $-Z(f)$ be taken for λ. The observed point must be adjusted to bring it into line with the model. Computer programs that fit the signal-detection model usually provide these corrected estimates of λ (or even give estimates of the bias) as part of their output. However, a calculational procedure is useful when the line has been sketched by eye or when published data are used. A good way to make this adjustment is to transfer the observed points to the closest point on the line. This adjustment is not completely optimal, as it does not take account of the sampling variability of f and h, but is close to the best solution. Figure 4.7 illustrates the process. The observed point $[Z(f),\, Z(h)]$ is translated to the point $(\tilde{z}_F,\, \tilde{z}_H)$ along a line perpendicular to the fitted line. The geometry here is identical to that used to determine the closest approach to the origin when finding d_a (Equation 4.3). In this case, it gives

$$\tilde{z}_F = \frac{b[Z(h) - a] + Z(f)}{1 + b^2} \qquad \text{and} \qquad \tilde{z}_H = a + b\tilde{z}_F. \tag{4.14}$$

These values are then substituted into Equation 4.13.

Example 4.4: Determine the bias parameters for the conditions in Figure 4.1.

Solution: The program that fitted the model in Example 4.1 gave estimates of the criteria. For the first condition, λ_1 was 1.017. Then, using Equation 4.12 with $a = 0.550$ and $b = 0.736$,

$$\log \beta_1 = \tfrac{1}{2}[\lambda_1^2 - (a - b\lambda_1)^2] + \log b$$

$$= \tfrac{1}{2}[(1.017)^2 - (0.736 \times 1.017 - 0.550)^2] + \log(0.736) = -0.191.$$

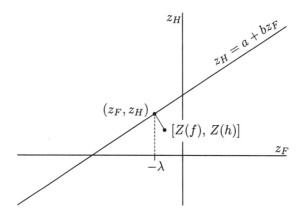

Figure 4.7: Shifting an observed point $[Z(f), Z(h)]$ to the perpendicularly nearest point $(\widetilde{z}_F, \widetilde{z}_H)$ on the fitted line to estimate the criterion or bias.

The other criteria are 0.518, 0.211, and -0.183, which give bias estimates of -0.186, -0.362, and -0.524, respectively.

If estimates of the criteria were not provided by the fitting program, then they need to be derived from the original point and the fitted line. For the first point, $f_1 = 0.165$ and $h_1 = 0.405$, so $Z(f_1) = -0.974$ and $Z(h_1) = -0.240$. Using Equation 4.14 to translate the observed point to the line gives

$$\widetilde{z}_{F_1} = \frac{b[Z(h_1) - a] + Z(f_1)}{1 + b^2}$$
$$= \frac{0.736(-0.240 - 0.550) + (-0.974)}{1 + (0.736)^2} = -1.009.$$

The criterion $\lambda_1 = 1.009$ obtained from this coordinate is essentially the same as the value 1.017 produced by the computer algorithm, the difference being due to the slightly different estimation criteria and to rounding error. The transformed hit rate corresponding to the new estimate is

$$\widetilde{z}_{H_1} = a + b\widetilde{z}_{F_1} = 0.550 + (0.736)(-1.009) = -0.193.$$

The point $(1.009, -0.193)$ falls on the line in Figure 4.1 (showing that a computational error has not been made). The bias estimate from Equation 4.13 is

$$\log \beta = \tfrac{1}{2}(\widetilde{z}_{F_1}^2 - \widetilde{z}_{H_1}^2) + \log b$$
$$= \tfrac{1}{2}[(-1.009)^2 - (0.193)^2] + \log(0.736) = 0.184.$$

Again, the small difference between this result and that of the program is unimportant.

4.6   Aggregation of detection statistics

The signal-detection measures in this chapter apply best to data obtained by collecting many observations from a single observer. In this way, good estimates of P_H and P_F can be made, and one can be reasonably confident that the underlying parameters are approximately stable. This type of data is frequently found in studies that are conducted in the psychophysical tradition from which signal-detection theory was developed. These studies typically use a small number of observers (sometimes just one), each of whom is analyzed separately and discussed individually.

In other domains, studies are common that use larger groups of observers, in this context usually called *subjects* or *participants*. In these studies, the researcher wishes to get some sort of average or mean result and draw inferences that apply to the group as a whole. For example, a memory researcher using a recognition paradigm may run 20 or more subjects in each of several conditions, then compare the conditions. A signal-detection analysis is necessary, because the researcher wishes to remove the response bias when making comparisons of recognition accuracy, particularly as it undoubtedly varies over subjects. The researcher takes each subject's rate of true and false recognition responses (i.e., hit rate and false-alarm rate) and calculates either d' or A_z. The data now must be aggregated over the subjects so that a single conclusion can be drawn.

Similar issues arise in a single-observer analysis when the data from a particular detection task are collected on a series of days. Separate sets of hit and false-alarm rates are obtained on each day. The researcher does not plan to discuss each day's data separately (even though they differ slightly), but wants a single measure to express the average performance.

There are two ways to pool signal-detection results over a set of entities, be they subjects or sessions. One possibility is to pool the observations over the individual entities to obtain the total number of trials, hits, false alarms, and so forth, then to run the analysis on these aggregate data. The other possibility is to conduct separate signal-detection analyses for each individual entity, then combine the results. The first procedure is the simpler to apply, but, when it is feasible, the latter approach is much to be preferred.

If the two approaches gave the same answer, then there would be no problem—either approach could be used. However, generally they give different answers. The more the entities vary from one to another, the more the procedures diverge. From a formal point of view, the problem is that the signal-detection statistics are nonlinear functions of the original data, and the average of a collection of nonlinear functions is not equal to that function applied to the average. Thus, a statistic such as d' that is calculated from the pooled hits and false alarms does not equal the average of the d'_i calculated on the individual entities. The difference is not large when the entities are very homogeneous in their detection properties, but can be substantial when the entities are heterogeneous.

Either approach to pooling the entities gives a valid answer for its particular form of combination. The proper measure to use is the one that best matches the researcher's intent. Almost always, the intent is to find the average value of the individual detection statistics. For example, the memory researcher knows that subjects differ in how well they remember material and in their biases. Pooling the data just blurs these differences. Furthermore, although there is no average subject whose performance is estimated by the pooled data, there is a mean over the population of subjects who might have been sampled for the study. Inferences about the situation are expressed in terms of this mean, and it is estimated by the average of the sample statistics, not by those derived from the pooled data. The same arguments apply to the psychophysical researcher who collects data from the same observer on 10 similar sessions. It is not surprising that the observer differs a bit from one session to the other, but these differences are unimportant, so the researcher is concerned with the average performance. Each researcher should separately calculate whatever index is important for each subject or session, then average them. Calculating individual measures has an ancillary advantage, in that it provides information about the extent to which the subjects or sessions vary. This information is necessary for the statistical analysis that is discussed in Section 11.7.

Example 4.5: Five subjects are given fifty trials on a recognition test in which half the items are old and half are new. The left-hand columns in Table 4.1 give the number of YES responses made by each subject to new and old items. What is the average recognition performance for these subjects?

Solution: The first step is to calculate the detection statistics for each individual subject. Dividing the frequencies of OLD responses by 25 gives the false-alarm and hit rates, and a standard equal-variance analysis gives the detection statistics in the final columns. These values are averaged to give the mean values of $\overline{d'} = 0.64$ and $\overline{A}_z = 0.67$.

Pooling the responses before applying the model is less appropriate. Doing so gives mean response rates of

$$\overline{f} = \frac{70}{125} = 0.560 \quad \text{and} \quad \overline{h} = \frac{93}{125} = 0.744.$$

	New	Old	f	h	$\widehat{d'}$	$\widehat{A_z}$
S_1	10	15	0.40	0.60	0.507	0.640
S_2	12	20	0.48	0.80	1.095	0.781
S_3	14	19	0.52	0.76	0.792	0.712
S_4	16	23	0.64	0.92	1.047	0.770
S_5	18	16	0.72	0.64	-0.224	0.437
Mean					0.643	0.668

Table 4.1: *Recognition responses for new and old items from 5 subjects on a 50-item recognition test with equal numbers of old and new items. See Example 4.5.*

Detection statistics derived from these values are $d' = 0.505$ and $A_z = 0.639$, somewhat less than the values obtained by the preferred method.

When the number of trials is small or the task particularly hard, some subjects may have a false-alarm rate that exceeds the hit rate. For example, subject S_5 in Example 4.5 has $f = 0.72$ and $h = 0.64$, for which the standard estimating equations give $\widehat{d'} = -0.22$. Presumably, d' for this subject is not actually negative, although it must be small—the best guess is $d' = 0$. The negative value occurred because sampling accidents happened to increase f and decrease h. It is tempting to replace the negative value by zero or to drop the subject from the analysis, but doing so biases the mean. Sampling accidents that went the other way, producing a too-large value of sensitivity would not be altered or censored. The net effect of systematically removing any negative deviations, is to inflate the sample mean relative to its population counterpart. Thus, it is important to keep the negative value in the analysis when averaging the detection statistics, as has been done in Example 4.5.

The use of individual detection statistics places demands on the study. There must be a sufficient number of observations available from each entity (subject or session) to make a reasonably accurate estimate of d', A_z, $\log \beta$, or whatever measure is desired. A researcher who is planning to use these techniques should ensure that enough data are collected in each condition to calculate stable values. For example, the tests given to the subjects in the memory study must include a sufficient number of items to give at least rough estimates of the individual hit and false-alarm rates. A study in which, say, five items of each type were used would be unsatisfactory.

Finding the average of the individual measures can encounter difficulties when the number of observations is small enough that false-alarm rates of $f = 0$ or hit rates of $h = 1$ occur. As noted in Section 2.3, the detection measures in such cases are indeterminate, and adjustments to give values for distance measures such as d' are arbitrary. When many such entities are present, it is usually advisable to use a bounded detection statistic such as A_z that is less affected by how the extreme scores are treated. One should also verify that any conclusions are not dependent on one's choice of adjustment procedure.

Reference notes

The material in this chapter draws on the same references that were cited for Chapter 2. For the material on the choice of measures, see Swets (1986b) and Swets and Pickett (1986). Examples of operating characteristics from a variety of domains are in Swets (1986a). The measure A' was first described by Pollack and Norman (1964); see Smith (1995) and Macmillan and Creelman (1996) for historical and theoretical discussion.

Exercises

4.1. The following data come from three detection experiments that use the same stimuli, but in which the proportion of signal and noise trials was varied to induce the observer to shift the criterion.

Signal proportion	f	h
High	0.04	0.40
Middle	0.31	0.70
Low	0.67	0.83

a. Plot these data and qualitatively evaluate the assumptions of the signal-detection model.
b. Sketch the distributions of noise and signal events and locate the criteria.
c. Estimate the four detection statistics Δm, d_e, d_a, and A_z.
d. Estimate the bias for each of the three conditions.

4.2. Find A_z for the data in Problem 2.6.

4.3. Calculate the detection statistics d_e, d_a, and A_z for the data in Problem 3.3. Estimate the bias for conditions A, D, and F.

4.4. A single observer performs simple detection of a signal in three conditions. In one condition, the observer is asked to minimize false alarms, in another to minimize misses, and in the third to roughly balance the two types of error. In each condition 200 signal trials and 200 noise trials are given. The results are as follows:

	Low F.A.			Balanced			Low miss	
	NO	YES		NO	YES		NO	YES
N	189	11	N	151	49	N	92	108
S	114	86	S	88	122	S	25	175

a. Which of the Gaussian models is the most reasonable choice to describe these data?

b. Estimate the area under the Gaussian operating characteristic.

c. Sketch the underlying distributions of evidence for the noise and signal conditions based on the model that you selected.

4.5. Suppose that the equal-variance Gaussian model with $d' = 1.5$ is a correct description of a single detection task. Determine the area A_z under the operating characteristic. Now consider a condition with $z_F = 0.2$. Find the associated hit rate from the Gaussian model and use z_F and z_H to calculate the trapezoidal area A_{trap}. Make similar calculations for conditions with $z_F = 0.5$ and $z_F = 0.75$. Compare the results to A_z. Illustrate what is going on with a diagram.

Chapter 5

Confidence ratings

The last two chapters have argued that the most robust measures of the detectability of a signal are those calculated from the isosensitivity contour. The largest difficulty in using them routinely is the demands that they place on the observer. Enough data must be collected at each of several levels of bias to get accurate estimates of the hit rate and the false-alarm rate. These data are not hard to obtain from a practiced observer who can devote the required time to the task and who can maintain a stable performance level as the bias is changed. They are harder to get from observers who have limited availability or from inexperienced observers who cannot maintain consistent performance throughout the study. These difficulties, for example, almost preclude using bias manipulations in the typical learning or memory experiment.

One way to retain the advantages of an analysis based on the isosensitivity contour while alleviating some of these problems is to increase the amount of data collected on each trial. That can be done by combining each detection response with an indication of the observer's confidence that the response is correct. The confidence is taken as providing evidence about where on the underlying decision axis the effect of that stimulus falls. By using these ratings, information that bears on several points of the isosensitivity contour can be collected on each trial. The analysis of such experiments is the topic of this chapter.

5.1 The rating experiment

A study that uses confidence ratings is conducted exactly like a yes/no detection study in all aspects except the response. In the rating task, each trial is made more informative by asking the observer to provide a

| | Response | | | | | |
	N3	N2	N1	Y1	Y2	Y3
Noise	166	161	138	128	63	43
Signal	47	65	66	92	136	294

Table 5.1: Three-level confidence ratings from a single practiced observer detecting a weak visual pattern. The data come from an experiment by L. A. Olzak and P. Kramer, and I thank Lynn Olzak for their use.

confidence rating along with the response. For example, an observer could follow the response of YES or NO by assigning one of three categories: SURE, UNCERTAIN, or GUESSING. The result is an ordered range of responses indicating increasing certainty about the presence of the signal. At one end of this rating scale is the NO-SURE response, then NO-UNCERTAIN and NO-GUESSING, then the YES responses starting with from YES-GUESSING and ending with YES-SURE. In effect, six response categories are obtained from the observer instead of two.

It is not necessary to collect the yes/no response and the confidence rating as two separate entities. One could equally well have the observer give a single number that ranges between 1, which indicates high confidence that no signal was present, and 6, which indicates high confidence that it was. The same six ordered levels are produced.

Table 5.1 shows an example of some data collected from a rating-scale experiment. In this study a well-practiced observer attempted to detect a faint regular brightness modulation of a uniform visual field (a CRT screen). The Noise condition corresponded to an unmodulated field; the Signal condition, to a pattern of faint vertical bands of light and dark strips produced by a sinusoidal brightness modulation along the horizontal axis of the display. A three-level rating scale, like the one just described, was conjoined with the yes/no decision to produced six categories. In Table 5.1 the detection responses are denoted by Y or N and the ratings by a number from 1 to 3.

A quick examination of Table 5.1 reveals three things. First, the response frequencies for the two stimuli clearly differ, so the signal was to some degree detectable. Second, the rating scale provides information beyond that in a response of YES or NO. An N3 rating is more likely to be from a noise trial than is an N1 rating. A similar gradient is present for the YES responses. Third, the observer's performance was imperfect. All types of responses were made to both stimuli, and even high-confidence NO responses to signals and high-confidence YES responses to noise are moderately frequent. This pattern is consistent with a moderately difficult detection task.

Although the data in Table 5.1 are substantially richer than simple yes/no data, they can be reduced to these two categories. Ignoring the rating, there are $166 + 161 + 138 = 465$ responses of NO to the blank field (noise) stimulus. Pooling the three levels of confidence for the other responses and for the signal stimulus gives a table classified only by YES and NO:

$$\begin{array}{c|cc} & \text{NO} & \text{YES} \\ \hline \text{Noise} & 465 & 234 \\ \text{Signal} & 178 & 522 \end{array} \qquad (5.1)$$

The detectability of the signal is evident in the greater number of YES responses to it than to the noise. A simple analysis like that in Chapter 2 gives $\widehat{d'} = 1.09$ with $\widehat{\lambda} = 0.43$. However, by collapsing this way, the grading of the rating data is lost. It is this graded character that gives the rating experiment its power.

5.2 The detection model for rating experiments

The analysis of rating-scale data is based on the same representation of the stimuli that was used to describe the yes/no experiment. There is a single evidence dimension X, and the observation of a stimulus produces a value on that dimension distributed according to the random variables X_n and X_s. In most applications, these random variables are given Gaussian distributions with different means and possibly different variances:

$$X_n \sim \mathcal{N}(0, 1) \quad \text{and} \quad X_s \sim \mathcal{N}(\mu_s, \sigma_s^2).$$

The only change in the model is in the response process, which must be able to describe the confidence ratings. Instead of a single criterion that separates the response NO from the response YES, there is a series of criteria that separate the confidence levels. Figure 5.1 shows the model for six-level responses. Five criterion lines divide the range of X into six parts. Each section of the abscissa is associated with a different compound response. The response to any value of X below the criterion λ_3 is NO and that to any value above λ_3 is YES. Within the NO range, the criteria λ_1 and λ_2 divide the confidence levels. Observations below λ_1 are given a rating of 3, those between λ_1 and λ_2 are given a rating of 2, and those between λ_2 and λ_3 are given a rating of 1. Similarly, the criteria λ_4 and λ_5 divide the

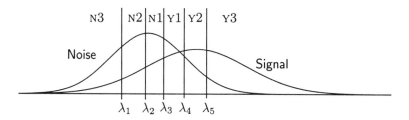

Figure 5.1: *Signal and noise distributions with five criteria separating the six response categories generated by three-level confidence ratings in a detection task.*

YES responses into three confidence levels. The most confident responses are made when large or small values of X are observed—those below λ_1 or above λ_5—and the least confident responses are made at intermediate values between λ_2 and λ_4.

This description is easily written in formal notation. Suppose that the rating experiment divides the responses into J categories. Let R be a discrete random variable that denotes the response that the observer makes. Thus, for Table 5.1, $J = 6$ and R takes values 1, 2, 3, 4, 5, or 6, corresponding to the responses from N3 through Y3. Assign end criteria that are very far to the left or right, $\lambda_0 = -\infty$ and $\lambda_J = \infty$. Now the probability of response $R = j$ is the area under the appropriate density function between the criteria λ_{j-1} and λ_j:

$$P(R{=}j|\text{noise}) = \int_{\lambda_{j-1}}^{\lambda_j} f_n(x)\, dx,$$

$$P(R{=}j|\text{signal}) = \int_{\lambda_{j-1}}^{\lambda_j} f_s(x)\, dx. \tag{5.2}$$

This representation differs from the one used for the yes/no task, only in that the integrals are bounded on both sides (compare Equations 5.2 to Equations 1.1 and 1.2).

For the Gaussian model, the integrals are replaced by differences in the Gaussian cumulative distribution function $\Phi(z)$. Using the parameters of the model or the slopes of the isosensitivity contour, Equations 5.2 become

$$P(R{=}j|\text{noise}) = \Phi(\lambda_j) - \Phi(\lambda_{j-1}), \tag{5.3}$$

$$P(R{=}j|\text{signal}) = \Phi\left(\frac{\lambda_j - \mu'_s}{\sigma_s}\right) - \Phi\left(\frac{\lambda_{j-1} - \mu'_s}{\sigma_s}\right) \tag{5.4}$$

$$= \Phi(b\lambda_j - a) - \Phi(b\lambda_{j-1} - a). \tag{5.5}$$

From these equations, one can calculate response probabilities for any set of parameters.

The collection of criteria in Figure 5.1 suggests one way to analyze the experiment. The two-by-two table of Display 5.1 was obtained by pooling the responses corresponding to the areas on either side of the criterion λ_3. Responses N3 through N1 were combined in the NO category, and responses Y1 through Y3 were combined in the YES category. Similar two-by-two tables are created by pooling the observations on either side of any other division between categories. In this way, five two-by-two tables are obtained from the six-level categorization, one corresponding to each criterion in Figure 5.1. Table 5.2 shows these tables, with their hit rates and false-alarm rates. One can then treat these five tables as if they were the results of five different bias manipulations, giving five (f, h) pairs. These points are plotted as a five-point operating characteristic in Figure 5.2, using both probability and Gaussian coordinates.

Two facts immediately emerge from an inspection of Figure 5.2. First, the transformed points fall very close to a straight line, which implies that a Gaussian model is satisfactory. Second, the slope of this line is less than $45°$, which implies that the unequal-variance model is necessary and that $\sigma_s^2 > 1$. The inadequacy of the unequal-variance model is also evident in the asymmetry of the probability operating characteristic.

The points on the isosensitivity contour obtained from a rating-scale study differ from those created by bias manipulations in several respects. Each point in a bias study derives from its own set of observations, while those in a rating study all come from the same observations. This difference has several consequences. The most obvious of these is that the rating-scale observations usually appear more stable than those derived from separate conditions. The way that the rating points are created by cumulating responses means that the empirical operating characteristic cannot have the sort of random nonmonotonic leaps and drops that appeared with the bias data in Figure 3.7 on page 48. It is easier to fit a straight line to rating data than to bias data, particularly by eye. Another nice property of rating data is that the points are not differentially affected by practice or fatigue effects. Even when there is some drift in the observer's performance, it affects all the points equally, not just those recorded later in the study.

The common origin of the points in the rating isosensitivity contour does mean that they are not statistically independent. For example, if it happened that the number of noise stimuli given N3 responses was unusually high in a particular study, then all points on the contour are pushed to the right. In contrast, the points of a bias-manipulated operating characteristic are independent. The principal effect of this dependence is technical: it must be accommodated in the algorithms used to fit the model. It also

	N3	N2-Y3	
Noise	166	533	$f = 0.762$
Signal	47	653	$h = 0.933$

	N3-N2	N1-Y3	
Noise	327	372	$f = 0.532$
Signal	122	588	$h = 0.840$

	N3-N1	Y1-Y3	
Noise	465	234	$f = 0.335$
Signal	178	522	$h = 0.746$

	N3-Y1	Y2-Y3	
Noise	593	106	$f = 0.152$
Signal	270	430	$h = 0.614$

	N3-Y2	Y3	
Noise	656	43	$f = 0.061$
Signal	406	294	$h = 0.420$

Table 5.2: Five two-by-two tables created by collapsing the rating data in Table 5.1.

increases the sampling variability of the parameter estimates over those for an operating characteristic based on bias manipulation with the same number of observations per point (although, of course, the bias study will require more observations overall).

5.3   Fitting the rating model

Because the underlying representation of the stimuli as two distributions is the same for the yes/no and the rating-scale models, the same measures of detectability are used with each. All the discussion in Chapter 4 applies. When the observations are regular enough to fit an operating characteristic by eye, then the measures are calculated as before. A computerized fitting algorithm for rating-scale data must take account of the fact that the points are not independent of each other, as they would be for a bias manipulation.

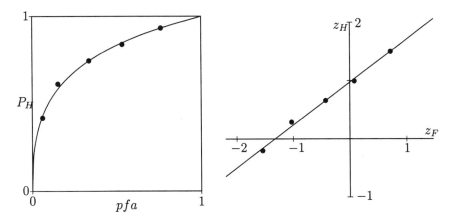

Figure 5.2: *Points corresponding to the five two-by-two tables in Table 5.2 that were created by collapsing the data of Table 5.1. The isosensitivity contours are derived from the Gaussian model fitted below.*

Programs of both types are available, and one should be sure to use the correct program for one's type of data.

Table 5.3 shows how the data in Table 5.1 is fitted in a form that accommodates the structure of a rating-scale study. The original response frequencies are entered in the columns labeled x_{nj} and x_{sj}. The frequencies of ratings higher than the category are cumulated in the columns $\sum x_{nk}$ and $\sum x_{sk}$ (the limits on these sums are from $j+1$ to 6). These entries are easily built up from the bottom: each is obtained by summing the two entries on the line below. So, $63 + 43 = 106$ and $128 + 106 = 234$. At the top of the table, this procedure gives the total number of trials of that type. In the next columns of Table 5.3, these sums are divided by the total frequencies to give proportions corresponding to false alarms f_j and hits h_j. These calculations repeat those in Table 5.2 in more compact form. The proportions are then converted to Gaussian coordinates, which are the values plotted in Figure 5.2. Besides being more compact, a display like Table 5.3 makes clear the dependence of the rows on each other, which the two-by-two configurations of Table 5.2 do not.

Example 5.1: Fit the rating model to the data in Table 5.1.

Solution: The transformed points in Figure 5.2 fall so close to a straight line that a fit by eye is nearly as accurate as a computer program. The straight-line isosensitivity contour in Gaussian coordinates is

$$z_H = 0.735 z_F + 0.974.$$

The criteria are calculated from the abscissas of this line at the fitted

		Noise				Signal		
j	x_{nj}	$\sum x_{nk}$	f_j	$Z(f_j)$	x_{sj}	$\sum x_{sk}$	h_j	$Z(h_j)$
		699				700		
1	166	533	0.762	0.714	47	653	0.933	1.498
2	161	372	0.532	0.081	65	588	0.840	0.994
3	138	234	0.335	−0.427	66	522	0.746	0.661
4	128	106	0.152	−1.030	92	430	0.614	0.290
5	63	43	0.061	−1.542	136	294	0.420	−0.202
6	43				294			

Table 5.3: Calculation of Gaussian isosensitivity coordinates for the data in Table 5.1.

points, as in Figure 4.7 on page 77. The five criterion values, λ_1 to λ_5, are

$$-0.714, \qquad -0.067, \qquad 0.423, \qquad 0.979, \qquad \text{and} \qquad 1.590.$$

From here on, the analysis is like that for a set of independent conditions. Following Equations 3.5, the intercept and slope are converted to the parameters of the Gaussian model:

$$\widehat{\mu}_s = \frac{a}{b} = \frac{0.974}{0.735} = 1.325,$$

$$\widehat{\sigma}_s = \frac{1}{b} = \frac{1}{0.735} = 1.360.$$

The four summary detection measures discussed in Chapter 4 are

$$\Delta m = \widehat{\mu}_s = \frac{a}{b} = 1.325,$$

$$d_e = \frac{2a}{1+b} = 1.123,$$

$$d_a = \frac{\sqrt{2}\,a}{\sqrt{1+b^2}} = 1.110,$$

$$A_z = \Phi\!\left(\frac{a}{\sqrt{1+b^2}}\right) = \Phi(d_a/\sqrt{2}) = \Phi(0.784) = 0.784$$

(by happenstance, the value of $\Phi(z)$ and its argument are identical here). These statistics can be used to compare this result to those measured in other conditions.

When the rating scale has more than three levels, there are more free observations in the data than are required to estimate the parameters. Each observed point of the isosensitivity contour (or each line of Table 5.3) reflects two data values. So, the six-level distributions in Table 5.1, which yield five points, have 10 free values. There are seven parameters in the model, μ_s, σ_s^2, and the five criteria λ_1 through λ_5. The excess of data over parameters makes it possible to test the fit of the model. Both the adequacy of the Gaussian assumption and the need for unequal variance can be investigated. These tests are described in Section 11.5 (and specifically for these data in Examples 11.8 on page 214 and 11.9 on page 218). They show that, up to the limits of the power of the test, the description of these data by the Gaussian model is satisfactory.

Exercises

5.1. A rating experiment is run with 150 noise trials and the same number of signal trials. A five-level rating scale is used for the responses. The numbers of responses of each type are

	1	2	3	4	5
Noise	42	49	20	24	15
Signal	10	20	15	39	66

a. Plot the operating characteristic, fit a line to it by eye.
b. Using this line, decide whether the Gaussian model acceptable. Can the variances of X_n and X_s be assumed to be the same?
c. Estimate the area A_z under the Gaussian operating characteristic.
d. If you have a computer program for fitting rating data, apply it and compare the results to your estimates.

5.2. A recognition memory study is run as follows. Subjects read a short (two page) story and are asked questions about its content. After half an hour of irrelevant activity, they are given a sheet containing 80 words. Half of these words had appeared in the story; the other half had not been used in any part of the experiment. The subjects are asked to circle those words that they remembered having been used and to put a check mark next to those words (either circled or not) for which they were very sure of their answer. The study involved various between-subject conditions, but these are not the concern of this problem, which deals with how to summarize each subject's data. The responses of one subject in this study are

	NO √	NO	YES	YES √	
New	15	13	7	5	40
Old	4	6	7	23	40

a. Plot the operating characteristics implied by these data.

b. Do these data support the use of a Gaussian model? Specifically, the equal-variance model?

c. Summarize each subject's recognition performance by a number between 0 and 1.

Chapter 6

Forced-choice procedures

The experiments described thus far in this book all use a single-stimulus detection procedure. One stimulus is presented on each trial, and the observer judges whether that stimulus contains a particular signal (and perhaps provides a confidence rating). Another way to measure detectability is by contrasting two stimuli on each trial. One of these stimuli is the signal; the other is noise. The observer's task is to identify which stimulus is the signal. Tasks of this type are known as *forced-choice procedures*, in this *two-alternative forced choice*. More generally, when the task involves one signal and $K - 1$ noise stimuli, it is a K-*alternative forced-choice procedure*. This chapter describes the analysis of data from these designs.

The essential feature of the forced-choice procedure is that the stimuli are contrasted within a single trial. In an ordinary single-stimulus detection task, with or without confidence ratings, each stimulus is judged in isolation; in a forced-choice task the observer can compare a signal stimulus to one or more noise stimuli. This explicit comparison changes the way that the task is described and analyzed. In a sense that is explained in Section 6.4, the comparison gives the observer an advantage over the observer in the single-stimulus task.

6.1 The forced-choice experiment

The simplest forced-choice experiment is one in which two alternatives are presented, and the observer tries to select the one that contains the signal. Consider how the tone-detection experiment described in Section 1.1 could be redesigned as a forced-choice study. As before, the observer listens to white noise throughout a trial. The indicator light is used to delimit two intervals, one following the other—it might go on at the start of the first

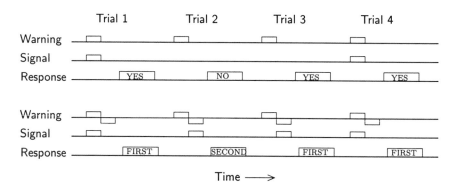

Figure 6.1: *The sequence of events in yes/no (top three lines) and forced-choice (bottom three lines) detection tasks. Four trials are shown, of which the third is an error.*

interval, blink between the first and the second intervals, and go off at the end of the second interval. The tone is added to the noise background in one of the two intervals, half the time in the first and half the time in the second. The observer has two buttons to press, one to indicate that the signal appeared in the first interval, the other that it appeared in the second interval. On each trial one of the buttons is pressed. The observer knows that there is a signal in one of the intervals and must select one of them, even when no signal is apparent.

Figure 6.1 contrasts the sequence of events in the yes/no and forced-choice procedures. The upper set of three lines shows four trials of a simple detection experiment. The top line shows the warning interval, which identifies the period when the signal might or might not be presented. The second line shows the signal, which here appears only on the first and fourth trials. The third line indicates the observer's YES or NO response. Trials 1, 2, and 4 are correct and trial 3 is a false alarm. The bottom set of lines shows the sequence of events in a forced-choice experiment with temporally separated stimuli. The warning signal now has two parts, indicated in the figure by excursions up and down; in the actual study these might be such things as a pair of tones, a pair of lights, one light that changes color or blinks between stimuli, or an arrow that points up or down. A signal appears on every trial, sometimes in the first interval (trials 1 and 4) and sometimes in the second (trial 2 and 3). The response line indicates whether the observer chose the first or second interval. Again, trials 1, 2, and 4 are correct and trial 3 is an error.

Converting a word-recognition experiment into forced-choice format is even more natural. In the first phase of the experiment the subjects are read a list of words to remember. After some intervening activity, the test phase commences. Each subject is given a sheet of paper on which are printed a series of pairs of words. Each pair consists of one old word and one new word, in either order. The subject is instructed to circle the old word in each pair. If neither word in a pair seems familiar, then the subject guesses. Each pair of words in this study constitutes a forced-choice trial.

The K-alternative version of these procedures simply adds more noise stimuli. A three-alternative version of the sensory task involves three intervals, only one of which contains the signal. A five-alternative memory test is created by presenting sets of one old word and four new words. The familiar multiple-choice test is a form of forced-choice procedure, at least when the examinee is required to answer every item.

The forced-choice procedure should be clearly distinguished from a second procedure, known as an *identification task*, with which it superficially shares some characteristics and with which it is sometimes confused. Suppose that there are K different types of signal, for example, tones of K different frequencies or material learned in K different contexts. In a K-alternative identification task the observer is presented with a single stimulus on each trial and asked to assign it to one of the stimulus types. As in the forced-choice task, one alternative from among K possibilities must be selected. However, in the forced-choice task K different stimuli are presented, while in the identification task only one stimulus appears. This difference means that the tasks are interpreted in very different ways. The identification task will be discussed in Chapter 7.

The trials in a forced-choice experiment do not differ among themselves as much as the trials in a yes/no experiment. Instead of some noise trials and some signal trials, each trial is composed of the same two (or K) events, merely rearranged. This homogeneity lets the results be scored very simply. The most natural measure of performance for a forced-choice task is the proportion of correct responses:

$$c = \frac{\text{Number of correct responses}}{\text{Number of trials}}.$$

When the signal is undetectable, the choice among the K alternatives is independent of the signal's location, and so $c \approx 1/K$. For detectable signals, the proportion of correct responses exceeds $1/K$, and for a highly detectable signal, c is close to one.

Example 6.1: In a two-alternative forced-choice study, 200 trials are presented to the observer. On a randomly selected half of these trials the signal is in the first interval and on half it is the second interval. On the 100 trials with the signal in the first interval, the observer chooses that first interval as the one containing the signal 84 times, and on the 100 trials with the signal in the second interval, the observer chooses the first interval 25 times. Give a measure of the observer's performance.

Solution: There were 84 correct responses when the signal came first and $100 - 25 = 75$ when it came second. The combined proportion of correct responses is

$$c = \frac{84 + 75}{200} = 0.795.$$

This number should be compared to the chance probability of $1/2$ that would be obtained by an observer who guessed randomly.

6.2   The two-alternative forced-choice model

A theoretical treatment of the two-alternative forced-choice procedure is based on the same probabilistic representation of a stimulus that was used with simple detection data. Any stimulus produces some unobserved sensation whose value is described by the random variable X. This variable has the form X_n for noise events and X_s for signal events. Each forced-choice trial supplies the observer with two realizations of these random variables, one realization of X_n and one realization of X_s. These observations are used to decide which stimulus was the signal.

The simplest model for the forced-choice task has the observer make a direct comparison of the observations and pick the interval that contains the larger value. This rule leads to a correct response whenever the realization of X_s exceeds that of X_n and to an error whenever the realization of X_n exceeds that of X_s. The correct-response probability is

$$P_C = P(X_s > X_n)$$
$$= P(X_s - X_n > 0). \tag{6.1}$$

This probability depends only on the distributions of X_s and X_n.

It is easy to derive the correct-response probability P_C under the Gaussian model. The two observations have Gaussian distributions with different means and possibly with different variances:

$$X_n \sim \mathcal{N}(0, 1) \quad \text{and} \quad X_s \sim \mathcal{N}(\mu_s, \sigma_s^2)$$

(Figure 6.2, top). Assume that these observations are independent. The difference of two independent Gaussian random variables is also Gaussian with a mean equal to the difference in the component means and a variance

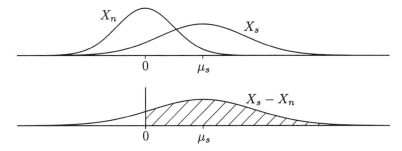

Figure 6.2: The distributions of X_n and X_s and the distribution of their difference in the Gaussian model for a two-alternative forced-choice task. The shaded area in the bottom panel corresponds to P_C.

equal to the sum (not the difference!) of the variances (Equation A.33 on page 235), so that

$$X_s - X_n \sim \mathcal{N}(\mu_s, 1 + \sigma_s^2)$$

(Figure 6.2, bottom). The probability that this random variable is positive is

$$P_C = 1 - \Phi\left(\frac{0 - \mu_s}{\sqrt{1+\sigma_s^2}}\right) = \Phi\left(\frac{\mu_s}{\sqrt{1+\sigma_s^2}}\right). \tag{6.2}$$

Example 6.2: Suppose that $\mu_s = 2$ and $\sigma_s = 1.5$. What is P_C?

Solution: These distributions are the ones shown in Figure 6.2. Using Equation 6.2,

$$P_C = \Phi\left(\frac{2}{\sqrt{1 + (1.5)^2}}\right) = \Phi\left(\frac{2}{1.802}\right) = \Phi(1.109) = 0.866.$$

The observer is correct on almost seven trials out of eight.

Under the equal-variance Gaussian model, $\mu_s = d'$ and $\sigma_s^2 = 1$, so that the relationship between the correct response probability and d' is

$$P_C = \Phi(d'/\sqrt{2}) \quad \text{and} \quad d' = \sqrt{2}\, Z(P_C). \tag{6.3}$$

The analogous equations are used with estimates of d' and the observed proportion c of correct responses:

$$c = \Phi(\widehat{d'}/\sqrt{2}) \quad \text{and} \quad \widehat{d'} = \sqrt{2}\, Z(c). \tag{6.4}$$

Example 6.3: Estimate how detectable the stimuli from the experiment in Example 6.1 would be when used in a yes/no task.

Solution: The correct-response rate in that study was $c = 0.795$. Thus, by the second member of Equations 6.4,

$$d' = \sqrt{2}\,Z(0.795) = (1.414)(0.824) = 1.165.$$

On the face of it, a substantial advantage of the forced-choice procedure is that it eliminates the need to consider the bias for or against reporting a signal. With a yes/no task, the observer may have unequal propensities for making YES and NO responses, often for the sensible reasons suggested by the discussion of optimal performance in Section 2.5. The decision criterion λ accommodates this bias. When the observer is required to make a choice between two alternatives, this type of bias is no longer an issue, for every trial contains a stimulus of each type. In effect, both a YES and a NO response is required on each trial. At the expense of a somewhat more complicated procedure, the scoring and interpretation are simplified.

6.3   Position bias

By using the forced-choice procedure, one avoids the problem of bias for or against detection of the signal. However, the procedure does not eliminate another type of bias. The stimuli presented during a forced-choice trial must always be separated in some way or other, temporally or spatially. It may be that the observer prefers one of these alternatives to the other. For example, an observer in the tone-detection experiment might be more inclined to assign the signal to the first interval than to the second. These biases are referred to as *temporal biases* or *position biases* as the case may be—the latter term will be used here for either type. When position bias exists, the analysis of the previous section must be expanded and its results qualified.

On the face of it, the symmetry of the forced-choice situation makes a position bias less likely than a bias toward YES or NO responses in the single-stimulus detection designs. In a well-designed forced-choice experiment, the signal occurs equally often in each location, and there is no differential reward, so that a rational observer will not prefer one alternative to the other. Nevertheless, there are several reasons that the stimuli can be treated asymmetrically and that a position bias can arise. One possibility is that the two stimuli interact. So, the first presentation in a temporally separated pair of stimuli may act to enhance or inhibit the second presentation. Another possibility arises from the order in which the observer examines the stimuli. An observer may systematically examine one position before the other—this ordering is inevitable with temporally separated stimuli, but can occur with spatially separated ones as well—and pay attention to the second position only when a decision cannot be made based on the first stimulus. Finally, the observer may prefer one response to the other for some idiosyncratic reason. For example, the right-hand response key may be easier to press than the left-hand key, creating a bias toward its use. However it arises, position bias cannot be ruled out, so it is important to understand how it affects performance.

The presence of a position bias changes the probability of a correct response from the bias-free value calculated in the last section. This probability P_C is maximum when the observer is unbiased, and it declines with the degree of asymmetry in the responses. In the extreme case of an observer who always selects the same position, only half the responses are correct, regardless of the separation between X_s and X_n.

The effects of a position bias are analyzed by recasting the analysis. Think of the pair of stimuli presented on a forced-choice trial as a single compound stimulus and the observer's task as the discrimination between these compounds. More specifically, in a two-alternative forced-choice design, the "stimulus" is one of two types. Either the signal comes first, temporally or by some spatial ordering, and the noise comes second, or the noise comes first and the signal second. Denote these trial types by $\langle SN \rangle$ and $\langle NS \rangle$, respectively. An observer who determines which of these events occurred also identifies the interval or location that contains the signal. Reorganized in this way, the data form a two-by-two table of stimulus by response. In the experiment described in Example 6.1, this configuration is

		Interval chosen	
		FIRST	SECOND
Stimulus	$\langle SN \rangle$	84	16
	$\langle NS \rangle$	25	75

This reorganized table is analogous to a detection table. It can be analyzed in the same way to give the usual statistics, such as $\widehat{d}'_{FC}$, $\widehat{\lambda}_{FC}$, and $\widehat{\log \beta}_{FC}$. The subscript FC here distinguishes these quantities from their counterparts in an ordinary yes/no detection task.

To make sense of these forced-choice statistics, one returns to the underlying signal-detection model. Assume that the observer notes the evidence favoring a signal at each position and compares them to see which is larger. Let X_1 be the observation made in the first position (it will be drawn from either X_n or X_s), X_2 be that made in the second position, and $Y = X_1 - X_2$ be the difference between them. On $\langle SN \rangle$ trials, X_1 has the signal distribution X_s, and X_2 has the noise distribution X_n, so the mean

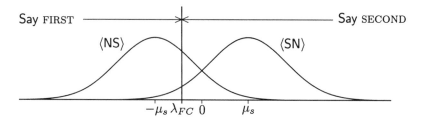

Figure 6.3: *The distribution of the difference* $Y = X_1 - X_2$ *for the two stimulus orders in a forced-choice task. The decision criterion shown creates a bias toward the first interval.*

of the distribution of the difference between them is positive:

$$Y_{\langle SN \rangle} = X_s - X_n \sim \mathcal{N}(\mu_s,\ 1+\sigma_s^2).$$

On $\langle NS \rangle$ trials, the assignment is reversed, and the distribution has a negative mean:

$$Y_{\langle NS \rangle} = X_n - X_s \sim \mathcal{N}(-\mu_s,\ 1+\sigma_s^2).$$

Figure 6.3 shows these distributions. A criterion λ_{FC} lets the observer separate these two random variables and identify which of the stimuli contains the signal. The decision rule is

$$\begin{cases} \text{If } Y \geq \lambda_{FC}, \text{ then say FIRST,} \\ \text{If } Y < \lambda_{FC}, \text{ then say SECOND.} \end{cases}$$

Because $Y_{\langle SN \rangle}$ and $Y_{\langle NS \rangle}$ are differences of the same two random variables, they have identical variances. The distance between their means expressed in standard deviation units is

$$d'_{FC} = \frac{\text{Difference between means}}{\text{Standard deviation of distribution}} = \frac{2\mu_s}{\sqrt{1+\sigma_s^2}}. \qquad (6.5)$$

The use of the equal-variance model here is not a consequence of any assumptions about the overall form of the noise and signal distributions, but of the fact that both $\langle SN \rangle$ and $\langle NS \rangle$ trials contain one stimulus of each type. The variance of Y is the same for the two trial types unless there are position-by-stimulus interactions that affect the variance (i.e., the variance depends interactively on both the stimulus and the position in which it appeared). Such effects are somewhat far-fetched in most designs. A convenient feature of the forced-choice procedure is that it allows performance to be summarized by a single number, even in situations where variance differences between the raw stimulus distributions are likely.

The representation illustrated in Figure 6.3 is like that of an equal-variance model for a yes/no task using a centered criterion (compare it to Figure 2.3 on page 27). Thus, the same equations can be used to estimate the differences between mean, the criterion, and the bias. Adapting the estimators from that discussion (Equations 2.4, 2.6, and 2.11) gives

$$\widehat{d}_{FC} = Z[\widehat{P}(\text{FIRST}|\langle\text{SN}\rangle)] - Z[\widehat{P}(\text{FIRST}|\langle\text{NS}\rangle)], \tag{6.6}$$

$$\widehat{\lambda}_{FC} = -\tfrac{1}{2}\{Z[\widehat{P}(\text{FIRST}|\langle\text{NS}\rangle)] + Z[\widehat{P}(\text{FIRST}|\langle\text{SN}\rangle)]\}, \tag{6.7}$$

$$\widehat{\log\beta}_{FC} = \tfrac{1}{2}\{Z^2[\widehat{P}(\text{FIRST}|\langle\text{NS}\rangle)] - Z^2[\widehat{P}(\text{FIRST}|\langle\text{SN}\rangle)]\}. \tag{6.8}$$

The bias values are positive when the second stimulus is preferred.

Although Equations 6.6–6.8 can be used to estimate the parameters of the forced-choice model, it is somewhat arbitrary to treat one of the stimulus configurations as if it were a "signal" and the other as if it were "noise." A more symmetrical, and more appealing, approach is to write the estimation equations using the correct-response probabilities:

$$P_{C\langle\text{SN}\rangle} = P(\text{FIRST}|\langle\text{SN}\rangle) \quad \text{and} \quad P_{C\langle\text{NS}\rangle} = P(\text{SECOND}|\langle\text{NS}\rangle).$$

The distance between means is the sum of the distance from each mean to the criterion. As a look at the areas in Figure 6.3 shows, the distance from $-\mu_s$ to λ_{FC} in standard deviation units is $Z(P_{C\langle\text{NS}\rangle})$ and the distance from λ_{FC} to $+\mu_s$ is $Z(P_{C\langle\text{SN}\rangle})$. Thus,

$$d'_{FC} = Z(P_{C\langle\text{SN}\rangle}) + Z(P_{C\langle\text{NS}\rangle}). \tag{6.9}$$

This equation, and comparable ones for the bias, gives the estimation equations

$$\widehat{d}_{FC} = Z(\widehat{P}_{C\langle\text{SN}\rangle}) + Z(\widehat{P}_{C\langle\text{NS}\rangle}), \tag{6.10}$$

$$\widehat{\lambda}_{FC} = \tfrac{1}{2}[Z(\widehat{P}_{C\langle\text{NS}\rangle}) - Z(\widehat{P}_{C\langle\text{SN}\rangle})], \tag{6.11}$$

$$\widehat{\log\beta}_{FC} = \tfrac{1}{2}[Z^2(\widehat{P}_{C\langle\text{NS}\rangle}) - Z^2(\widehat{P}_{C\langle\text{SN}\rangle})]. \tag{6.12}$$

As in the analysis of yes/no task, $\log\beta_{FC} = d'_{FC}\lambda_{FC}$.

Example 6.4: Find the detection statistics for the data in Example 6.1. *Solution:* The proportions of correct choices for the $\langle\text{NS}\rangle$ and $\langle\text{SN}\rangle$ conditions are

$$\widehat{P}_{C\langle\text{SN}\rangle} = {}^{84}\!/_{100} = 0.84 \quad \text{and} \quad \widehat{P}_{C\langle\text{NS}\rangle} = {}^{75}\!/_{100} = 0.75.$$

The separation between the distributions is

$$\widehat{d}_{FC} = Z(0.84) + Z(0.75) = 0.674 + 0.994 = 1.67.$$

Both the criterion placement and the bias are negative:

$$\widehat{\lambda}_{FC} = \tfrac{1}{2}[Z(0.75) - Z(0.84)] = \tfrac{1}{2}[0.674 - 0.994] = -0.16$$

and

$$\widehat{\log \beta}_{FC} = \tfrac{1}{2}[Z^2(0.75) - Z^2(0.84)] = \tfrac{1}{2}[(0.674)^2 - (0.994)^2] = -0.27.$$

The negative values imply a shift of the criterion toward the left and an apparent preference for FIRST responses.

The overall probability P_C of a correct response is a combination of the individual correct-response probabilities, $P_{C\langle SN \rangle}$ and $P_{C\langle NS \rangle}$, weighted by the probabilities of the trial types, $P_{\langle SN \rangle}$ and $P_{\langle NS \rangle}$ (the law of total probability, Equation A.5 on page 227):

$$P_C = P_{\langle SN \rangle} P_{C\langle SN \rangle} + P_{\langle NS \rangle} P_{C\langle NS \rangle}.$$

Varying the criterion changes the probability. For the Gaussian model,

$$
\begin{aligned}
P_C &= P_{\langle SN \rangle} P(Y_{\langle SN \rangle} < \lambda_{FC}) + P_{\langle NS \rangle} P(Y_{\langle NS \rangle} > \lambda_{FC}) \\
&= P_{\langle SN \rangle} \Phi\left(\frac{\lambda_{FC} - (-\mu_s)}{\sqrt{1+\sigma_s^2}}\right) + P_{\langle NS \rangle} \left[1 - \Phi\left(\frac{\lambda_{FC} - \mu_s}{\sqrt{1+\sigma_s^2}}\right)\right] \\
&= P_{\langle SN \rangle} \Phi\left(\frac{\mu_s + \lambda_{FC}}{\sqrt{1+\sigma_s^2}}\right) + P_{\langle NS \rangle} \Phi\left(\frac{\mu_s - \lambda_{FC}}{\sqrt{1+\sigma_s^2}}\right).
\end{aligned}
\tag{6.13}
$$

When there is no position bias (i.e., $\lambda_{FC} = 0$), the two Gaussian terms are the same, and this result is identical to the value derived without considering position bias (Equation 6.2). When $P_{\langle SN \rangle} = P_{\langle NS \rangle} = \frac{1}{2}$, as it is in most studies, optimal performance occurs when $\lambda_{FC} = 0$. When the trial types are not equally frequent, it behooves the observer to shift the criterion to increase the correct-response probability for the most frequent type (see Problem 6.6).

Some unevenness is natural in any set of observed frequencies, so it is useful to have a way to tell whether any bias apparently present in a set of data should be taken seriously. A test of the hypothesis of no bias is made by comparing the actual data to the expected frequencies calculated under the assumption of no bias. Without position bias, the probability of a correct response for the two types of trials should be the same. This common probability is estimated by the observed correct-response proportion c. On average, the number of correct responses to $\langle SN \rangle$ trials is the product of this proportion and the number of trials, $N_{\langle SN \rangle}$. With similar products for the other three cells, the table of frequencies describing a lack of position bias is

	FIRST	SECOND
$\langle SN \rangle$	$cN_{\langle SN \rangle}$	$(1-c)N_{\langle SN \rangle}$
$\langle NS \rangle$	$(1-c)N_{\langle NS \rangle}$	$cN_{\langle NS \rangle}$

One now determines how much the data deviate from these values by calculating the Pearson statistic X^2 (Equation A.54 on page 246). There are four data observations, two constraints on the data ($N_{\langle SN \rangle}$, $N_{\langle NS \rangle}$), and the value of c to be estimated, so the test has $4 - 2 - 1 = 1$ degree of freedom (Equation A.56). Values of X^2 that are large in relation to the values from chi-square distribution in the table on page 251 imply the presence of position bias.

Example 6.5: Investigate whether the data in Example 6.1 show a position bias.

Solution: The observed correct-response rate is

$$c = \frac{75 + 84}{200} = 0.795.$$

With 100 trials in each row, the expected frequencies under the assumption of equal bias are

	FIRST	SECOND
$\langle SN \rangle$	79.5	21.5
$\langle NS \rangle$	21.5	79.5

The Pearson statistic comparing these values to the observed data is

$$X^2 = \frac{(25-21.5)^2}{21.5} + \frac{(75-79.5)^2}{79.5} + \frac{(84-79.5)^2}{79.5} + \frac{(16-21.5)^2}{21.5} = 2.39.$$

The deviation of the data from the equal-bias model as measured by this statistic does not exceed the critical value of 3.84 obtained from the chi-square table on page 251 (using the 5% level and one degree of freedom). The observed asymmetry is not sufficient to reject the hypothesis of no bias. The slight favoring of the first position seen in the data is consistent with what one might find in an unbiased observer.

6.4 Forced-choice and yes/no detection tasks

The detectability of a particular signal can be investigated with either a forced-choice or a yes/no experiment. Measures that are based directly on parameters of the underlying model, such as the values of d' from Equations 6.3 and 6.4, should be similar if the assumptions of the model are satisfied. However, measures such as P_C and d'_{FC} in the forced-choice procedure do not have immediate counterparts in simple detection. The signal-detection model provides the structure needed to establish a correspondence between the tasks.

The relationship between the simple detection and forced-choice procedures is expressed by two remarkably simple relationships. The first of these relates d' measured in a simple detection study to d'_{FC} measured in a forced-choice study. It applies to an observer who is operating by the equal-variance Gaussian model and has no position bias. The link is through the correct-response probability P_C. Equations 6.3 and 6.9 link the two distance measures to this probability:

$$d' = \sqrt{2}\, Z(P_C) \qquad \text{and} \qquad d'_{FC} = 2Z(P_C)$$

(remember that P_C does not depend on which order the stimuli are presented for an observer who has no position bias). Combining the two relationships by eliminating $Z(P_C)$ gives

$$d'_{FC} = \sqrt{2}\, d'. \tag{6.14}$$

Thus, the forced-choice measure d'_{FC} is about 41% larger than the d' measured in a simple detection study.

One should not interpret this result as showing that detection of the stimulus is better in a forced-choice task than in a yes/no task. The difference only reflects the fact that the observer in a forced-choice task has more information on which to base a response than does the observer in a simple detection task. The first observer sees two stimuli, the second only one, and two observations are better than one.

The second relationship is far more general. It asserts that, whatever distributions are associated with the stimuli, the unbiased observer's probability of a correct response in a two-alternative forced-choice task equals the area under that observer's isosensitivity contour obtained from yes/no detection:

$$P_C = A. \tag{6.15}$$

This simple, but very important, relationship is known as the *area theorem*. As mentioned in Section 4.3, it gives a ready interpretation to the area as a measure of detectability. It is a remarkable fact that the area theorem does not depend on the specific form of the noise and the signal distributions (such as having a Gaussian distribution), but is true whatever the form of these distributions.

When a Gaussian distribution is assumed, the area theorem connects the theoretical area A_z under the isosensitivity curve and the parameters of the model. Using the correct-response probability when there is no position bias (Equation 6.2), the area theorem implies that

$$A_z = P_C = \Phi\left(\frac{\mu_s}{\sqrt{1+\sigma_s^2}}\right).$$

(6.16)

This form of the area theorem was used in Section 4.3 to express the area in terms of the parameters of the Gaussian model (Equation 4.6).

The remainder of this section proves the area theorem. It is not necessary to follow this proof (which is more difficult than most of the derivations in this book) to understand or use the result. However, the theorem is sufficiently important that some readers will find its proof satisfying; those who do not can skip to the next section.

The theorem is proved by finding general expressions for the two sides of Equation 6.15, one for P_C and one for A. These expressions turn out to be the same.

First consider P_C in the two-alternative forced-choice design. A correct response is made when the value of the signal event X_s exceeds that of the noise event X_n. If the noise event happens to take the value u, then the probability of this event is

$$P(\text{correct}|X_n{=}u) = P(X_s > u) = \int_u^\infty f_s(x)\,dx = H(u),$$

where $H(u)$ is the hit function defined in Section 3.2 (Equations 3.1 on page 45). To find the unconditional probability of a correct response, accumulate the conditional correct-response probability over the distribution of X_n, weighting each value $H(u)$ by how likely that value of u is to occur, as specified by the density function $f_n(u)$. Mathematically, the accumulation is an integral over the range of X_n of the conditional probability of a correct response times the weighting density:

$$P_C = \int_{-\infty}^\infty H(u)f_n(u)\,du.$$

(6.17)

This equation is the continuous form of the law of total probability (Equation A.10 on page 228).

Now turn to the area under the isosensitivity curve for a detection test. In terms of the hit and false-alarm functions $H(u)$ and $F(u)$, the operating characteristic is $R(p) = H[F^{-1}(p)]$ (Equation 3.2 on page 45), and the area under it is

$$A = \int_0^1 R(p)\, dp = \int_0^1 H[F^{-1}(p)]\, dp$$

(Equation 4.5). To evaluate this integral, make the substitution[1] $u = F^{-1}(x)$ or $x = F(u)$. Because $F(u)$ is the integral of $f_n(x)$ above u (Equations 3.1), its derivative is $F'(u) = -f_n(u)$. So dp is replaced by $-f_n(u)\, du$, giving

$$A = \int_{F^{-1}(0)}^{F^{-1}(1)} H(u)[-f_n(u)\, du] = -\int_{\infty}^{-\infty} H(u) f_n(u)\, du.$$

The minus sign is removed by changing the direction of the integral, so that it goes from $-\infty$ to ∞. The result is identical to the right-hand side of Equation 6.17. Thus, A and P_C must be equal, completing the proof of the area theorem.

6.5   The K-alternative forced-choice procedure

In the two-alternative forced-choice task, the observer chooses between one alternative that contains a signal and one that does not. This procedure readily generalizes to designs in which several alternatives are presented. In the K-alternative forced-choice procedure, one signal and $K - 1$ noise events are randomly assigned to K positions or intervals. The observer must report which position contains the signal. The number of noise events makes this task more difficult when K is large than when it is small.

The signal-detection model for K-alternative forced-choice data is similar to that for the two-alternative case. The information available to the observer on a trial is expressed by one realization of the signal random variable X_s and $K - 1$ independent realizations of the noise random variable, $X_{n1}, X_{n2}, \ldots, X_{n,K-1}$. The observer without position bias chooses the alternative associated with the largest value. Thus, for a correct response to be made, the realization of X_s on that trial must exceed every realization

---

[1] To replace the variable x in an integral $\int f(x)\, dx$ by the new variable $u = g(x)$, the argument x is replaced by $g^{-1}(u)$, the differential dx by $du/g'(x)$, and the limits a and b by the functions $g(a)$ and $g(b)$.

of X_n. The probability of a correct response is the probability of this event:

$$P_C = P[X_s > \max(X_{n1}, X_{n2}, \ldots, X_{n,K-1})]. \tag{6.18}$$

When $K = 2$, there is only a single noise event, and Equation 6.18 is just the probability that $X_s > X_n$ (Equation 6.1). For larger values of K, one of the noise events is more and more likely to exceed the single realization of X_s and cause an error to be made.

The evaluation of Equation 6.18 draws on results from the theory of order statistics (discussed in advanced texts) and is similar to the derivation of Equation 6.17 in the proof of the area theorem. Suppose first that the value of X_s is known. For a correct response to occur when $X_s = x$, this value must exceed all $K - 1$ observations of X_n. The probability that any particular X_n is less than x is given by the cumulative distribution function $F_n(x)$, and the probability that all $K - 1$ independent realizations of X_n are less than x is the product of these $K - 1$ individual values:

$$P(\text{correct response}|X_s=x) = P[\max(X_{n1}, X_{n2}, \ldots, X_{n,K-1}) < x]$$
$$= [F_n(x)]^{K-1} = F_n^{K-1}(x)$$

(the last step is notational). To get the unconditional probability of a correct response, these values are averaged over the distribution of signal events. For each x, the correct-response probability is weighted by the density function $f_s(x)$ and the products are accumulated in an integral:

$$P_C = \int_{-\infty}^{\infty} P[\max(X_{n1}, X_{n2}, \ldots, X_{n,K-1}) < x] f_s(x)\, dx$$
$$= \int_{-\infty}^{\infty} F_n^{K-1}(x) f_s(x)\, dx. \tag{6.19}$$

This integral (like that in Equation 6.17) should be thought of as an average, not an area. It states that P_C is the average probability that signal exceeds the noise observations, this average being taken over the distribution of signal strengths.[2] For any particular distributions of X_n and X_s, Equation

---

[2]Many seemingly different versions of this formula are found. In some treatments, the probability of an error is found instead of that of a correct response. To make an error, one of the $K - 1$ noise events must exceed the signal event and the remaining $K - 2$ noise events. Integrating over the distribution of noise events and multiplying by the number of noise events, any of which could be the largest, gives the probability

$$P(\text{error}) = (K - 1) \int_{-\infty}^{\infty} F_s(x) F_n^{K-2}(x) f_n(x)\, dx.$$

Although this integral and Equation 6.19 look superficially different, they can be shown to be equivalent. The hit function and the false-alarm function can be used instead of the cumulative distribution functions.

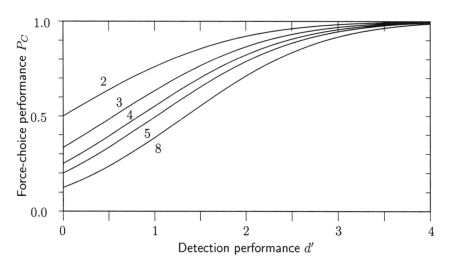

Figure 6.4: *Correct-response probabilities P_C for the equal-variance Gaussian forced-choice model with various numbers of distractors as a function of detection performance d'. The upper line for $K = 2$ is identical to the plot in Figure 4.4.*

6.19 can be evaluated, numerically if not otherwise, to find the correct-response probability.

For the Gaussian model, there is no closed-form solution of Equation 6.19. However, numerical integration is not difficult for a computer, and routines to evaluate these equations are easily prepared. Table 6.1 gives the correct-response probabilities under the equal-variance Gaussian model for K-alternative forced-choice designs with K between 2 and 9. Rows of the table give different amounts of separation between the signal and noise distributions, as measured by d'. A selection of these values are plotted in Figure 6.4. As this figure shows, P_C increases with the distance between distributions and decreases with the number of noise stimuli. These probabilities provide a satisfactory representation of the forced-choice experiments and a way to translate parameters between the different tasks.

Example 6.6: In a four-alternative forced-choice experiment a correct response rate of 60% was obtained. What level of performance can be expected in a two-alternative forced-choice task and a yes/no detection task with the same stimuli under the equal-variance Gaussian model?

Solution: In Table 6.1, a correct-response rate of 0.6 with $K = 4$ falls midway between the rows labeled 1.1 and 1.2. Thus $d' \approx 1.15$, which is the value of d' that would be expected in the simple detection task. This value can also be obtained from Figure 6.4. Start at the point 0.6 on the

	Number of alternatives K							
d'	2	3	4	5	6	7	8	9
0.0	0.500	0.333	0.250	0.200	0.167	0.143	0.125	0.111
0.1	0.528	0.362	0.276	0.224	0.189	0.163	0.144	0.128
0.2	0.556	0.391	0.304	0.250	0.212	0.185	0.164	0.148
0.3	0.584	0.421	0.333	0.277	0.238	0.209	0.186	0.169
0.4	0.611	0.452	0.363	0.305	0.264	0.234	0.210	0.191
0.5	0.638	0.483	0.393	0.335	0.293	0.261	0.236	0.216
0.6	0.664	0.513	0.425	0.365	0.322	0.289	0.263	0.242
0.7	0.690	0.544	0.456	0.396	0.353	0.319	0.292	0.270
0.8	0.714	0.574	0.488	0.429	0.384	0.350	0.322	0.299
0.9	0.738	0.604	0.520	0.461	0.417	0.382	0.353	0.330
1.0	0.760	0.634	0.552	0.494	0.449	0.414	0.385	0.361
1.1	0.782	0.662	0.583	0.526	0.483	0.447	0.418	0.394
1.2	0.802	0.690	0.614	0.559	0.516	0.481	0.452	0.427
1.3	0.821	0.716	0.644	0.591	0.549	0.514	0.486	0.461
1.4	0.839	0.742	0.674	0.622	0.581	0.548	0.520	0.495
1.5	0.856	0.766	0.702	0.653	0.614	0.581	0.553	0.530
1.6	0.871	0.789	0.729	0.682	0.645	0.613	0.587	0.563
1.7	0.885	0.810	0.754	0.711	0.675	0.645	0.619	0.597
1.8	0.898	0.830	0.779	0.738	0.704	0.676	0.651	0.629
1.9	0.910	0.849	0.802	0.764	0.732	0.705	0.682	0.661
2.0	0.921	0.866	0.823	0.788	0.758	0.733	0.711	0.691
2.1	0.931	0.882	0.843	0.811	0.783	0.760	0.739	0.720
2.2	0.940	0.896	0.861	0.832	0.807	0.785	0.766	0.748
2.3	0.948	0.909	0.877	0.851	0.828	0.808	0.790	0.774
2.4	0.955	0.921	0.893	0.869	0.848	0.830	0.814	0.799
2.5	0.961	0.931	0.907	0.885	0.867	0.850	0.835	0.822
2.6	0.967	0.941	0.919	0.900	0.883	0.869	0.855	0.843
2.7	0.972	0.949	0.930	0.913	0.899	0.885	0.873	0.862
2.8	0.976	0.957	0.940	0.925	0.912	0.900	0.890	0.880
2.9	0.980	0.963	0.949	0.936	0.924	0.914	0.904	0.896
3.0	0.983	0.969	0.956	0.945	0.935	0.926	0.918	0.910
3.2	0.988	0.978	0.969	0.961	0.953	0.946	0.940	0.934
3.4	0.992	0.985	0.978	0.972	0.967	0.962	0.957	0.953
3.6	0.995	0.990	0.985	0.981	0.977	0.974	0.970	0.967
3.8	0.996	0.993	0.990	0.987	0.985	0.982	0.980	0.977
4.0	0.998	0.995	0.993	0.992	0.990	0.988	0.986	0.985
4.5	0.999	0.999	0.998	0.997	0.997	0.996	0.995	0.995
5.0	1.000	1.000	0.999	0.999	0.999	0.999	0.999	0.998

Table 6.1: Values of d' and of P_C for K-alternative forced choice.

ordinate, proceed horizontally until the $K = 4$ line is reached, then drop down to read $d' = 1.15$ on the abscissa.

One way to find the expected two-alternative forced-choice performance is to use Equation 6.3:

$$P_C = \Phi(d'/\sqrt{2}) = \Phi(1.15/1.414) = \Phi(0.833) = 0.79.$$

Another possibility is to interpolate between two rows of Table 6.1 to get the same value. A third is to use Figure 6.4: start at $d' = 1.15$ on the abscissa, go up to the $K = 2$ line, then over to read off the ordinate of 0.79. The area theorem says that this value is also the area under the isosensitivity contour.

When using the relationships between tasks shown in Table 6.1 and Figure 6.4, one should remember that they are predicated on the assumption that the equal-variance Gaussian model holds and the observer has no position bias. If that model is inapplicable, usually because $\sigma_s^2 \neq 1$ or the observer prefers one position over the others, then the correspondence is inaccurate.

One sometimes sees applications of the forced-choice model to situations in which the number of alternatives is very large, such as in the hundreds or thousands. Such uses push the model too far. Before taking the values obtained from Equation 6.19 seriously in these applications, it is important to consider what they imply. As K increases, the probability that all noise stimuli fall under a given realization of X_s decreases, and, for any value of μ_s, the probability P_C falls to zero as K goes to infinity. To get a reasonably large value of P_C when K is very big, the distributions of X_s and X_n must be placed quite far apart. When that is done, the distribution of the maximum of the noise events depends critically on the exact shape of the upper tail of $f_n(x)$. Since the Gaussian distributions are idealizations, such a heavy reliance on this aspect of their form is unwarranted. If the tails of the noise distribution do not have exactly the form specified by the theoretical model, then the value of P_C obtained from Equation 6.19 is in error, often seriously so. This sensitivity to distributional extremes contrasts sharply with the relative robustness of the central statistics. A small violation of Gaussian form has a negligible effect on the utility of the transformation of f and h to $\widehat{\lambda}$ and $\widehat{d'}$ in ordinary detection or in the analysis of a two-alternative forced-choice design, but it has a big effect on the calculation of P_c in a many-alternative forced-choice design. Thus, although the K-alternative forced-choice model may reasonably be used to describe the selection of one correct word from a set of four words, it is not an acceptable model for the choice from a list of 50 words.

Exercises

6.1. Suppose that $\mu_s = 2$ and $\sigma_s^2 = 5$. What is the probability of a correct response in a forced-choice experiment if the observer is not biased toward either interval?

6.2. A 300-trial forced-choice experiment recorded the following responses:

<table>
<tr><td></td><td colspan="2" align="center">Interval chosen</td></tr>
<tr><td></td><td align="center">FIRST</td><td align="center">SECOND</td></tr>
<tr><td>⟨NS⟩</td><td align="center">52</td><td align="center">98</td></tr>
<tr><td>⟨SN⟩</td><td align="center">80</td><td align="center">70</td></tr>
</table>

a. Estimate the proportion of correct responses, the detection statistic d'_{FC}, and the bias.

b. Test the data for a lack of position bias.

6.3. Two stimuli are presented in the same location, one after the other. A small increment is added to the intensity of one of the stimuli, and the observer tries to identify which interval contains the increment. Correct responses are made 95 of 120 times when the increment is in the first interval and 75 of 120 times when it is in the second interval. The researcher hypothesizes that the presence of the first stimulus inhibits the response to the second stimulus, making an increment more difficult to detect in that interval. Is this hypothesis supported by these data?

6.4. In a yes/no detection task an observer has a hit rate of 0.72 and a false-alarm rate of 0.16.

a. Predict the probability that this observer makes a correct response in a two-alternative forced-choice task with the same signals.

b. What assumptions about the observer are necessary to make this prediction?

c. Predict performance in a five-alternative forced-choice task.

6.5. In the first phase of a study, subjects read a passage for its content. In a later phase, they are tested to see if they can recognize specific words from the passage. The word-recognition tests are conducted in two ways:

- Words pairs consisting of one old word and one new word are presented, and the subject is asked to circle the old word. The numbers of correct responses out of 50 items for 8 subjects are 39, 42, 45, 37, 47, 36, 40, and 40.
- Groups of five words consisting of one old word and four new words are presented, and again the old word is circled. Eight different subjects

are tested on the same 50 target words and receive correct-response scores of 34, 40, 36, 39, 38, 33, 38, and 34.

Compute recognition statistics for each subject and compare their average values (recall Section 4.6). Can difference in performance between the tasks be attributed to the difference in procedure, or is one type of test intrinsically easier than the other?

6.6. Suppose that the frequencies of $\langle SN \rangle$ and $\langle NS \rangle$ trials were unequal. Maximize Equation 6.13 to show that the optimal placement of the criterion is at

$$\lambda_{FC}^* = -\frac{1+\sigma^2}{2\mu_s} \log \frac{P_{\langle NS \rangle}}{P_{\langle SN \rangle}}.$$

Note: This problem requires the use of calculus.

Chapter 7

Discrimination and identification

The discussion so far has concerned the detection of a weak signal. The observer's goal is to determine whether this signal has occurred or in which of two or more intervals it fell. Signal-detection theory describes how the ambiguities of these weak signals affect the response. Another psychophysical task is known as *discrimination*. Here, one of two (or more) types of signals is presented on each trial. The signals have sufficient intensity that the observer can easily detect them. Instead of deciding whether or not a stimulus had occurred, the observer must decide which of the possible signals was presented. More generally, in an *identification* task, the observer is presented with one of I distinct stimuli, each of which has a known label. The observer assigns a label to the stimulus, trying to correctly identify it.

Consider three examples of discrimination tasks. The first is psychophysical, a pitch discrimination experiment. The stimuli are two pure tones of slightly different frequency. After having listened to the tones in isolation so that they are familiar, the observer is presented with a series of tones embedded in white noise. On each trial, the observer must decide whether the tone was at the higher or the lower of the two frequencies. Although the tones are loud enough that they can easily be detected, the two frequencies are very close together and are sufficiently masked by the noise that they are hard to identify. The second example involves memory. In the first phase of this study, the subject listens to a list of words, some read by a man and some by a woman. In the second phase, the subject is given a sheet of paper containing the words from the list and must identify, based on memory, which of the two speakers had said each word. The subject's task is not to decide if a word had been on the list—all the words on the sheet had been studied—but who said it. The third example is more complex. A radiologist is looking at abnormal X-ray films. One of several different abnormalities is present in every film, and the radiologist must assign it to its correct type. Some form of pathology—the signal—is present on each film, but its diagnosis is not known. The important characteristic of these examples is that the presence of a signal is not in doubt, but its nature is.

In some ways, the discrimination and identification tasks are very similar to the detection task, while in other ways, they differ from it. This chapter examines those similarities and differences. It emphasizes those parts of discrimination theory that are closely allied to detection theory.

7.1 The two-alternative discrimination task

In certain respects, discrimination and detection are much alike. In both tasks, the observer is presented with stimuli that are sufficiently similar to be confused, and in both the observer must classify them. In this sense, one can think of detection as discrimination between signal and noise. This similarity lets much of detection theory be carried over to analyze a discrimination experiment. Both tasks are modeled by positing a dimension that expresses the evidence for the two alternatives. Stimulus events correspond to observing a value along this continuum. Observations of the two stimuli have different distributions, largely because one is shifted along the evidence dimension relative to the other. Thus, a decision between them can be made by selecting a value along the continuum and classifying the stimuli according to the side of this criterion on which the observation falls. The theory of likelihood-ratio testing that will be described in Chapter 9 provides a theoretical basis for either task.

The formal similarity of detection and discrimination lets one apply techniques used in the analysis of the detection data to get measures of discrimination performance. One way to approach the analysis is to arbitrarily treat one of the signals as if it were a "noise" stimulus and the other signal as if it were "signal plus noise." The "false alarms" are incorrect assignments of the first stimulus as the second, and "hits" are correct identifications of the second stimulus. From these proportions, discrimination statistics such as $\widehat{d'}$, $\widehat{A_z}$, and $\widehat{\log \beta}$ can be estimated using the equations that apply to detection experiments or with standard computer programs. These quantities accurately summarize the performance, and can be used to compare conditions.

Although the numbers that come out of such an analysis are correct, the assignment of the stimuli to noise and signal events is conceptually awkward. Why should one tone be a signal and the other noise? It is

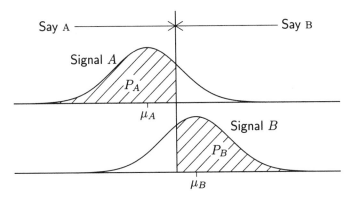

Figure 7.1: *The signal-detection model for two-alternative discrimination. The shaded areas correspond to correct responses.*

more attractive to treat the discrimination stimuli symmetrically. Instead of speaking of hits and false alarms, a better representation is to use the proportion of correct responses made to each stimulus. Denote the two signals by A and B and the responses by A and B. The probabilities of correct responses of each type are given by the conditional probabilities $P_A = P(\text{A}|A)$ and $P_B = P(\text{B}|B)$. Denote the corresponding proportions by their lowercase equivalents, p_A and p_B. Figure 7.1 shows the distributional model of the task. The decision axis is represented by a single dimension, and observing a stimulus gives a value along this axis. The distribution of the observations depends on the which stimulus was presented. A decision criterion divides the continuum into response regions associated with the two responses. The shaded areas in the figure correspond to the probabilities P_A and P_B of correctly identifying the two types of stimulus.

The decision model of Figure 7.1 also implies the existence of a discrimination operating characteristic. By varying the position of the decision criterion, performance can be varied from an emphasis on A responses, with $P_A \approx 1$ and $P_B \approx 0$, to an emphasis on B responses, with $P_A \approx 0$ and $P_B \approx 1$. The relationship of the two probabilities is shown in Figure 7.2. This discrimination operating characteristic is reversed left to right from the detection operating characteristics because the horizontal axis plots correct responses instead of errors. Its interpretation is the same. In particular, the area under the curve has the same meaning as a measure of the distinguishability of the signals.

Suppose further that the distributions are Gaussian, with means μ_A and μ_B and common variance σ^2. Because the decision continuum cannot be directly observed, exact values for μ_A, μ_B, σ^2, and the decision criterion

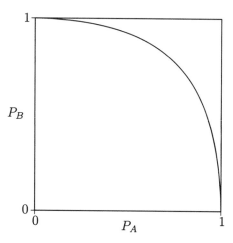

Figure 7.2: *The discrimination operating characteristic for the distributions in Figure 7.1.*

cannot be determined without an additional constraints. It is necessary to fix both the center and the scale of the picture before unique parameters can be determined. The scale is fixed by setting the variance σ^2 to unity. The zero point of the scale is also arbitrary, and unlike the detection case, there is no reason to single out one distribution to be the center. However, the most relevant performance parameters of this model do not depend on where the origin is set. Neither the distance between the means of the distributions, $d' = \mu_B - \mu_A$, nor the likelihood-ratio bias $\log \beta$ depends on the placement of the origin. These quantities can be found without having to commit oneself to an origin.

The parameters of the discrimination model are estimated by adapting the comparable detection equations to the new context. An estimate of the distance between the means of the distributions is found from the Z transforms of the correct-response probabilities (adapting Equation 2.4 on page 24):

$$\widehat{d'} = Z(p_A) + Z(p_B). \tag{7.1}$$

The second term is added here rather than subtracted because correct responses are measured instead of the false alarms of the basic detection model. The area under the discrimination operating characteristic is obtained from Equation 4.8 on page 68:

$$\widehat{A_z} = \Phi(\widehat{d'}/\sqrt{2}). \tag{7.2}$$

This area has the same interpretation as the correct-response probability in a two-alternative forced-choice experiment that it does for detection. Following Equation 2.11 on page 30, an estimate of the bias, ordered so that positive values indicate a preference for the observer to choose alternative A, is

$$\widehat{\log \beta} = \tfrac{1}{2}[Z^2(p_A) - Z^2(p_B)]. \tag{7.3}$$

Similar equations were constructed in Chapter 6 for the model for forced choice with position bias (Equations 6.10–6.12 on page 101).

Example 7.1: Suppose that in the speaker's-voice experiment, 50 words were used, 25 in each voice. A subject correctly identifies 20 of speaker A's words and 15 of speaker B's words. Describe the performance.

Solution: The proportions of correct responses are $p_A = 0.8$ and $p_B = 0.6$. From Equation 7.1, the discriminability is

$$\widehat{d'} = Z(p_A) + Z(p_B) = Z(0.8) + Z(0.6) = 0.842 + 0.253 = 1.095.$$

The area under the operating characteristic (Equation 7.2) is

$$\widehat{A_z} = \Phi(\widehat{d'}/\sqrt{2}) = \Phi(1.095/1.414) = \Phi(0.774) = 0.78.$$

Equation 7.3 gives the bias:

$$\widehat{\log \beta} = \tfrac{1}{2}[Z^2(p_A) - Z^2(p_B)] = \tfrac{1}{2}[(0.842)^2 - (0.253)^2] = 0.322.$$

The positive value of $\widehat{\log \beta}$ indicates a preference to choose speaker A, a characteristic that is evident from the fact that $p_A > p_B$.

The discussion of optimal detection performance in Section 2.5 applies to discrimination tasks with few changes. For the equal-variance model, the optimal criterion lies midway between the means, where $\log \beta = 0$. If the two types of signal are unequally likely, then the criterion is shifted away from the more common signal, and the optimal bias equals the logit of the stimulus probabilities (Equation 2.15 on page 34):

$$\log \beta^\star = \log \frac{P(\text{Stimulus } A)}{P(\text{Stimulus } B)}. \tag{7.4}$$

Although the basic signal-detection representations of detection and discrimination are very similar, there are some important differences between them. Most of these differences arise because the stimuli used in discrimination differ from each other in their qualities, whereas those used in detection differ in the presence or absence of the signal. The qualitative differences among the discriminative signals have two important consequences. The first consequence concerns the way that the differences between signals are represented. All the stimuli in a detection experiment fall on a single continuum of intensity that ranges from the complete absence of a signal to an intense signal. This variation is essentially unidimensional. In contrast, the stimuli in a discrimination task can differ from each other in a multitude of ways. Some stimuli have a single dimension that underlies their variation. For example, pure tones can be ordered on the frequency dimension (although even with pure tones the physical and psychological relationships are more complex, as any musician knows). With other stimuli, the stimulus structure is intrinsically multidimensional. Some multidimensional stimulus sets have a fixed dimensionality; examples are colored patches that differ along the three dimensions of hue, saturation, and brightness. In other sets the number of dimensions is large and uncontrolled, as in the speaker's voice or X-ray example. Some issues related to this multidimensionality are discussed in Section 7.3.

Another consequence of the differences between the tasks concerns the plausibility of the equal-variance model. In a detection experiment there is an intrinsic asymmetry between the two types of stimuli. A noise stimulus lacks a quality that is present in the signal stimulus. This asymmetry allows their distributional properties to differ. In particular, one may have good reasons, such as those mentioned in Section 3.4, to believe that the signal distribution has a larger or smaller variance than the noise distribution. Such an asymmetry is less likely in many discrimination tasks, particularly those with well-controlled stimuli. The stimuli to be discriminated have nearly the same intensity and differ only by a small amount in some other aspect. Because differences in intensity are negligible, the distributional properties, and particularly the variance, are usually nearly the same.

7.2   The relationship between detection and discrimination

A discrimination experiment can be run either with weak signals that are hard to detect or with strong signals that can be detected without difficulty. When weak signals are used, an extension of the signal-detection model provides a connection between the two tasks. Consider three stimuli, a noise stimulus N and two signal stimuli, A and B, each of which is weak enough so that detection is imperfect. Three tasks can be constructed from these stimuli: detection of A, detection of B, and discrimination between A and B. The models for the three tasks are linked by assuming that the

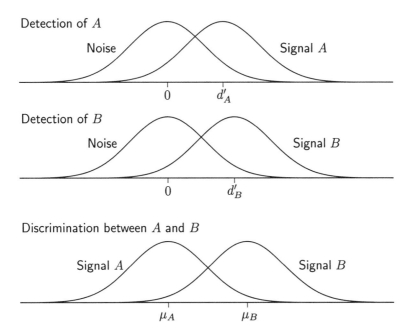

Figure 7.3: *Representations of two detection tasks and of discrimination between the two signals on separate unidimensional decision axes.*

same distribution applies to a given stimulus when it is used in different tasks. The tasks differ in the axes along which the decisions are made.

To be specific, suppose that the two detection conditions give the statistics $d'_A = 1.5$ and $d'_B = 1.8$, and that the discrimination performance is $d'_{AB} = 2.13$. The three pairwise signal-detection models for these results are shown in Figure 7.3. An integrated picture should combine these three separate panels. The six distributions in Figure 7.3 derive from only three stimuli, and ultimately they should refer to only three distributions. The center of the distribution for a particular stimulus should be the same regardless of the task—for example, the center of the A distribution at d'_A in the upper panel for the detection task should be the same as the center μ_A in the lower panel for the discrimination task. To put the three parts of Figure 7.3 together, the origins in the two top panels are mapped to a common point, as are the points d'_A and μ_A in the first and third panels and the points d'_B and μ_B in the final two panels. The three axes then form the triangle at the top of Figure 7.4. When the three means are aligned, the configuration is forced into two dimensions. In this form, it can represent both quantitative and qualitative differences among the stimuli. The magnitude or intensity of a stimulus is represented by its distance from the

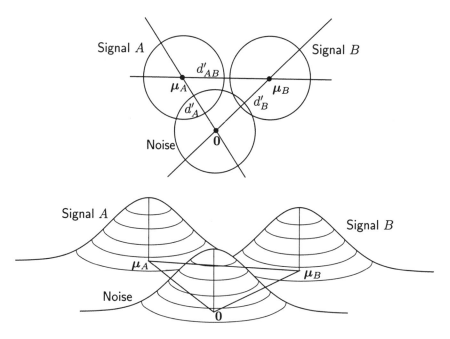

Figure 7.4: Detection and discrimination of weak signals. The upper panel shows two-dimensional configuration of the means created by combined the panels of Figure 7.3. The lower panel shows three bivariate density functions placed over these points.

origin, and the qualitative difference between two signals corresponds to the fact that they are displaced from the origin in different directions.

The distributions of the observations in this two-dimensional space are bivariate. A noise stimulus generates an observation from a random variable X_N with a mean at the origin—the boldface letter indicates a multivariate (here bivariate) quantity or vector. The two signals are represented by random variables X_A and X_B with means away from the origin 0 at two distinct points μ_A and μ_B. Figure 7.4 shows the distributions, at the top by isodensity contours drawn at one standard deviation from the mean (they enclose about 40% of the distribution), and at the bottom as bell-shaped hills (see the description of the multivariate Gaussian distribution on page 243). The assumption that $\sigma_n = 1$ determines the diameter of the circle about the origin. Under the equal-variance detection model, the signal circles have the same size as the noise circle. In an unequal-variance model, the diameters of the signal circles are larger or smaller than that of the noise circle. The three individual pictures in Figure 7.3 are created by

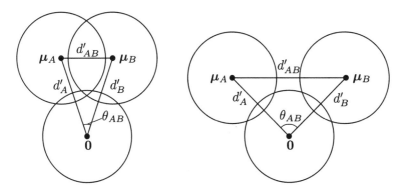

Figure 7.5: *Two-dimensional configurations representing equally detectable stimuli with similar qualities (left) and independent qualities (right).*

slicing through the density functions along vertical planes that pass through the lines that connect each pair of means.

In a diagram such as Figure 7.4, the distance between pairs of means relative to the size of the circles measures the detectability or discriminability of the two signals. The levels of performance for the three tasks are shown by the overlaps of these circles, in the same way that the level of performance is shown by the overlap of the distributions in the two-dimensional picture of Figure 7.3. The diagram also captures a second aspect of the relationship between the signals. When the two signals are qualitatively very similar, the means μ_A and μ_B are near each other and the angle θ_{AB} between the lines that connect them to the noise mean at $\mathbf{0}$ is small (Figure 7.5, left). When they are qualitatively different the angle is larger, and when they have completely unrelated qualities, θ_{AB} is a right angle (Figure 7.5, right). When two signals are identical in quality and cannot be discriminated, then $\theta_{AB} = 0$. The cosine of θ_{AB} makes a convenient measure of the signal similarity. Unrelated signals have $\cos\theta_{AB} = \cos 90° = 0$. Signals that share attributes have a smaller angle between the axes, so that $\theta_{AB} > 0$. Signals with identical qualities have $\cos\theta_{AB} = \cos 0° = 1$.

The angle θ_{AB} is calculated by combining information from the detection and discrimination statistics. The three d' statistics determine the lengths of the three sides of the triangle in Figure 7.4, and the law of cosines from trigonometry gives the angle between two sides. Applied to the model here, this rule states that

$$\cos\theta_{AB} = \frac{(d_A')^2 + (d_B')^2 - (d_{AB}')^2}{2d_A' d_B'}. \tag{7.5}$$

When θ_{AB} is known, Equation 7.5 can be solved for d_{AB}', giving an equation that relates the discriminability of signals to their detectability and degree of relatedness:

$$d'_{AB} = \sqrt{(d'_A)^2 + (d'_B)^2 - 2d'_A d'_B \cos\theta_{AB}}. \tag{7.6}$$

Example 7.2: Two signals are detected with d' statistics of 1.2 and 1.4, and they are discriminated with d'_{AB} of 1.6. Use the detection-discrimination model to predict the discrimination performance when the magnitudes of the signals are increased so that their detection statistics are 2.1 and 1.8.

Solution: First, determine the relationship between the signals. Using Equation 7.5, their similarity is

$$\widehat{\cos\theta}_{AB} = \frac{(\widehat{d'}_A)^2 + (\widehat{d'}_B)^2 - (\widehat{d'}_{AB})^2}{2\widehat{d'}_A \widehat{d'}_B} = \frac{(1.2)^2 + (1.4)^2 - (1.6)^2}{2 \times 1.2 \times 1.4} = 0.25.$$

The angular relationship between them is $\widehat{\theta}_{AB} = \arccos(0.25) = 75.5°$. The signals are not highly related, although (subject to sampling error) they are not perfectly unrelated. Carrying this angle over to the new experiment, the predicted discrimination statistic, using Equation 7.6 with the new values of d'_A and d'_B and the estimate $\widehat{\theta}_{AB}$, is

$$d'_{AB} = \sqrt{(d'_A)^2 + (d'_B)^2 - 2d'_A d'_B \cos\theta_{AB}}$$
$$= \sqrt{(2.1)^2 + (1.8)^2 - 2 \times 2.1 \times 1.8 \times 0.25} = 2.4.$$

When the signals have nothing in common, the situation is much simpler. The two stimulus dimensions are at right angles, and their discriminability can be calculated directly from their detectabilities. With $\cos\theta_{AB} = 0$, Equation 7.6 reduces to the root-mean-square average of the detectabilities:

$$d'_{AB} = \sqrt{(d'_A)^2 + (d'_B)^2}. \tag{7.7}$$

For signals of equal detectability, discrimination exceeds detection by a factor of $\sqrt{2}$:

$$d'_{AB} = \sqrt{2} d'. \tag{7.8}$$

When measured by d', discrimination performance is about 40% better than detection performance. This relationship connects detection and discrimination when detection of the two signals is mediated by different processes.

Example 7.3: A researcher believes that two signals stimulate different sets of physiological receptors. In the first phase of the experiment, the stimulus level that gives a detectability of $d' = 1.5$ is determined. What level of performance would be expected for a task in which these stimuli are discriminated?

Solution: With unrelated signals of equal strengths, the $\sqrt{2}$ rule applies. By Equation 7.8,

$$d'_{AB} = \sqrt{2}d' = 1.414 \times 1.5 = 2.1.$$

An observed value of d'_{AB} that is appreciably smaller than this might be taken as evidence that the researcher's assumption of independent receptors is wrong.

The relationships of detection and discrimination discussed here and the picture of Figure 7.4 involve several assumptions. Violations of these assumptions compromise its applicability. One concern involves the use of d' statistics. These are based on the assumption that the signal and noise distributions have equal variance. When this assumption is false, the relationships given by Equations 7.5 and 7.6 are no longer correct. Specifically, if one falsely assumed an equal-variance model when the signal variance is actually greater than the noise variance, then the value of d' estimated from the data would be less than the actual distance μ_s and the angle θ_{AB} would be underestimated. When $\sigma_n^2 > \sigma_s^2$, the opposite errors are made. In either case, the mistaken assumption can lead an investigator to incorrectly reject independence of the signals (see Problem 7.3).

When information about σ_s^2 is available, it can be incorporated in the measurement of the difference between signal means. For signals of equal detectability (as in Equation 7.8), the signal distribution must be scaled appropriately by replacing d' by μ_s/σ_s. Thus, the predicted discrimination of unrelated, equally detectable, signals is $d'_{AB} \approx \sqrt{2}\,\mu_s/\sigma_s$. The problems raised by unequal variance are more serious when more than one level of detectability is involved, either because the two signals are not equally detectable or in an attempt to extrapolate to a new stimulus levels. If the signal variance increases with the mean, then the distributions at the different levels have different variances and the equal-variance model for discrimination implicitly used to get Equation 7.6 is incorrect. The type of extrapolation over different signals that was done in Example 7.2 is likely to yield the wrong answer. The calculation is not useless, however. It provides a benchmark to which real performance can be compared. By trying to figure out what is wrong with the picture in Figure 7.4, one begins to understand what the actual observer is doing.

The relationships described in this section make sense only for weak signals. In many discrimination studies, stimuli are used that are well above the level where detectability is an issue at all. For example, one might determine the stimulus magnitude necessary for detection at $d' = 1$, then run a discrimination experiment using closely spaced stimuli with 10 times this magnitude. Choosing the stimulus magnitude as a multiple of the detection level ensures that the different signals have approximately comparable magnitudes. This procedure is particularly valuable in studies that involve several discrimination conditions run with different stimuli, where it is important that differences in magnitude cannot be used as cues. However, it breaks the connection between detection and discrimination. Extreme detection values, such as $d' \approx 10$, are well beyond the power of any experiment to measure. It makes no sense to apply probabilistic signal-detection theory to such "suprathreshold" stimuli at all.

7.3   Identification of several stimuli

The identification design generalizes to experiments in which more than two stimuli are to be discriminated. Before the start of these experiments, the observer learns the names of three or more types of stimuli. On each trial, a stimulus is presented and the observer tries to name it. For example, the signals might be tones of three distinct but readily confused frequencies. The observer hears a tone and classifies it as LOW, MEDIUM, or HIGH. An identification study can use more complex stimuli. In a study of the effectiveness of camouflage, the observer begins by studying pictures of vehicles—cars, types of trucks, and so forth. Then photographs of obscured or camouflaged vehicles are briefly presented, and the observer classifies them as one of the originally learned types.

The data from an identification experiment with I stimuli consists of I sets of responses, one for each signal. These responses are conveniently structured as a *confusion matrix*, such as Table 7.1, in which each row corresponds to a signal and each column to a response. The entries in the confusion matrix are the frequencies with which particular responses were made. From there, the conditional proportions of the responses given each stimulus are calculated. Entries along the diagonal of the confusion matrix describe correct responses, and off-diagonal entries describe the various types of errors.

Performance in an identification task can range anywhere from chance to perfect, and the confusion matrix has a characteristic form in these cases. When the stimuli are highly discriminable, most or all of the observations lie in the diagonal cells, and the off-diagonal entries are small or zero. When

		Raw frequencies						Proportions		
	A	B	C	D		A	B	C	D
A	139	25	21	15	A	0.695	0.125	0.105	0.075
B	91	35	29	45	B	0.455	0.175	0.145	0.225
C	51	16	41	92	C	0.255	0.080	0.205	0.460
D	25	17	24	134	D	0.125	0.085	0.120	0.670

Table 7.1: *Confusion matrices for an identification study with four stimuli, A, B, C, and D. Each row of the left-hand table gives the classification frequencies for 200 presentations of a stimulus. The right-hand table gives the conditional proportions.*

the stimuli cannot be discriminated, the proportion of observations in each column reflect only the observer's biases and are nearly the same in every row. The most useful data have an appreciable proportion of the entries on the diagonal, but various errors scattered off the diagonal, as in Table 7.1. Presumably, the most frequent confusions are between stimuli that are in some sense close to each other, and a model for the data should express this structure.

A wide variety of models have been developed to represent confusion data, and properly belong to the domain of *multidimensional scaling*. To explore the range of scaling models would take this discussion too far away from the topic of this book. However, certain of the multidimensional scaling models are worth noting because they derive naturally from the signal-detection representation.

Two-alternative detection theory can be extended to describe multialternative identification in several ways. An important distinction among these extensions concerns the dimensionality of the stimulus structure. Some stimulus sets are essentially unidimensional, and others are multidimensional. The set of tones of different frequencies constitute what might be a unidimensional set while the set of vehicle pictures is almost certainly multidimensional—it would be wise to check both of these assumptions empirically. Although similar models can be applied to both types of stimuli, the model for the univariate continuum is considerably simpler.

Start with the univariate stimulus set. The natural way to model what the observer is doing here is to extend the two-alternative discrimination model of Figure 7.1 by adding distributions for the new stimuli and cutpoints for the responses, as shown in Figure 7.6. Instead of the distributions of two stimulus random variables, there are now I distributions, one for each of the I stimuli. Instead of a single criterion dividing the continuum,

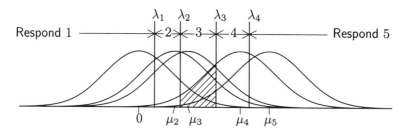

Figure 7.6: *The model for an I-item identification experiment with a unidimensional continuum. The shaded area gives the probability P_{23} of incorrectly identifying the third stimulus as the second.*

there are now $I - 1$ criteria separating the I responses. Denote the random variable associated with the ith stimulus by X_i and its density function by $f_i(x)$. In the most practical version of the model, X_i has a Gaussian distribution with unique mean μ_i and variance σ_i^2. To anchor the picture and fix its scale, give the lowest stimulus a standard distribution, so that $X_1 \sim \mathcal{N}(0, 1)$. The response criteria are located at positions $\lambda_1, \lambda_2, \dots,$ λ_{I-1}. When x falls below λ_1, response 1 is made; when it falls between λ_1 and λ_2, response 2 is made; and so on up to response I, which is made when $x > \lambda_{I-1}$.

Mathematically, the response probabilities are the integrals of the stimulus distributions over the various response ranges. Stimulus i gives rise to an observation of the random variable X_i with the density function $f_i(x)$. The probability P_{ij} that this stimulus is classified into category j is the area between criterion λ_{j-1} and criterion λ_j:

$$P_{ij} = \int_{\lambda_{j-1}}^{\lambda_j} f_i(x)\,dx. \tag{7.9}$$

To make this equation apply to all stimuli, including the first and the last, it is helpful to think of extreme criterion lines at $\lambda_0 = -\infty$ and $\lambda_I = \infty$. For the Gaussian model, these probabilities are

$$P_{ij} = \Phi\left(\frac{\lambda_j - \mu_i}{\sigma_i}\right) - \Phi\left(\frac{\lambda_{j-1} - \mu_i}{\sigma_i}\right). \tag{7.10}$$

For many sets of data, an equal-variance version of the model is appropriate. One can set $\sigma_i^2 = 1$, and Equation 7.10 becomes

$$P_{ij} = \Phi(\lambda_j - \mu_i) - \Phi(\lambda_{j-1} - \mu_i). \tag{7.11}$$

A comparison of this identification model to the rating-scale model of Chapter 5 (compare Figure 7.6 to Figure 5.1 on page 86) shows that formally they are nearly identical. Both pictures show a continuum with overlapping distributions along it. The axis is divided into regions associated with distinct responses by an ordered series of criteria. The equations giving the response probabilities are formally identical—Equations 5.2, 5.4, and 5.5 for the rating model and Equations 7.9, 7.10, and 7.11 for the identification model. The models differ in the number of stimuli and response categories. As the rating model was described, there were only two stimulus types, so the number of response categories exceeded the number of responses. In the identification model there are as many responses as stimuli.

The formal similarity of the identification and rating models means that similar estimation techniques can be used to fit them. When the ordering of the stimuli is known in advance (as it is for tones), the fitting algorithm is identical. A rating-scale estimation program that has been written to accommodate more than two stimuli may be used to fit the identification model. When the ordering of the stimuli is unknown, it must be inferred from the data. Some exploration is usually necessary to get the categories in the correct order. Until the ordering is correct, the program may fail in unexpected and unpleasant ways, as it can when the stimuli are not unidimensional.

Example 7.4: Assume that the data in Table 7.1 come from an ordered attribute, with their ordering as given in the table. Fit a unidimensional model to these data. Is the equal-variance assumption tenable?

Solution: The model was fitted using a rating-scale program. Under the equal-variance assumption, the program placed the four means at

$$0.00, \quad 0.62, \quad 1.25, \quad \text{and} \quad 1.76,$$

the first value being fixed. The criteria lie between these means, at

$$0.537, \quad 0.897, \quad \text{and} \quad 1.352.$$

A test of the goodness of fit of this model, as described at the end of Section 5.3, gives a Pearson statistic of $X^2 = 8.39$. With six degrees of freedom, the unidimensional model cannot be rejected (the critical value at the 5% level is 12.6).

To investigate the applicability of the equal-variance assumption, a model is fitted, again using the computer, that allows each stimulus to have a different variance. Naturally the fit is better (the model has more parameters), but it must be substantially so for the equal-variance model to be rejected. For these data, the improvement is not commensurate with number of added parameters. How much is gained can be measured and tested. The test statistic for a likelihood-ratio test, following the

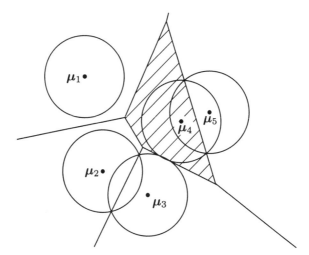

Figure 7.7: A two-dimensional representation of five stimuli in an identification experiment. Events in the shaded area are identified as instances of the fourth stimulus.

procedure discussed in Section 11.6 and illustrated in Example 11.9 on page 218, is $G^2 = 2.84$ on three degrees of freedom. Reference to a chi-square distribution shows that this value is not sufficient to reject the equal-variance model.

The unidimensional model illustrated in Figure 7.1 is quite restricted. For it to apply, there must be a single dimension that differentiates the signals. However, many sets of stimuli differ in several aspects, and an observer is able use some or all of them to make the discrimination. Such stimuli include the X-ray films mentioned at the start of this chapter and the camouflaged vehicles in this section. To model such stimuli, their means must be located in a multidimensional space, not confined to a line. Figure 7.7 illustrates a two-dimensional representation with $I = 5$ stimuli. Each stimulus is represented by a bivariate random variable, shown by a mean μ_i and a circular isodensity contour, as in Figure 7.4. Once each trial, when a stimulus is presented, the observer receives a realization x from the one of the random variables, but does not know which one. The observation must be assigned to one of the I categories. Figure 7.7 shows the simplest decision rule for this task: assign the observation to the signal distribution whose mean is closest to x. This rule has the effect of partitioning the space into a set of polygonal regions associated with the responses. In Figure 7.7, the region associated with the fourth response category is shaded.

In Figure 7.7, all the distributions have similar symmetrical forms and the observations are uncorrelated—their isodensity contours are circles of identical size. With more complex stimulus structures, the nearest-mean rule no longer works. A more general rule, which accommodates complex stimuli, can be developed from the likelihood-ratio principle that will be described in Chapter 9 and used in Chapter 10.

One consequence of a multidimensional characterization is that a simple ordering of the stimuli is no longer possible. In the unidimensional Figure 7.6, the means are ordered, with $\mu_2 < \mu_3 < \mu_4$, so that the third stimulus falls between the second and the fourth. In the bidimensional representation of Figure 7.7, these means have one ordering in the horizontal direction and a different one in the vertical direction.

Fitting the multidimensional model requires techniques that go beyond the scope of this book. Material on multidimensional scaling should be consulted for the fitting of this and other such models.

Reference notes

The relationship between detection and discrimination discussed in Section 7.1 was first described by Tanner (1956), although many of the ideas in other forms go back much farther. Multidimensional scaling is a field by itself. For an introduction, including the closest-distance metric representation mentioned here, see Kruskal and Wish (1978).

Exercises

7.1. An observer in an identification experiment receives 150 presentations of stimulus A and 200 presentations of stimulus B. The following assignments are made:

		Response	
		A	B
Stimulus	A	105	45
	B	37	163

a. Calculate the detection statistic A_z for these data.
b. Calculate the bias and compare it to that of an unbiased observer and that of an ideal observer who maximizes correct responses.

7.2. Detection data for two stimuli are

		F.A.	Hits
Stimulus	A	0.27	0.77
	B	0.17	0.79

What discrimination performance should be observed if the two stimuli are detected independently?

7.3. Suppose that an unequal variance model holds for the detection of two stimuli, with $\mu_s = 1.5$ and $\sigma_s^2 = 4.0$. A researcher performs detection and discrimination experiments with stimuli of this magnitude. However, the researcher does not check to see whether the unequal-variance model of detection is necessary and incorrectly uses an equal-variance analysis.

a. Assume that the two stimuli are detected from independent information. What detection performance d'_{AB} would be observed?

b. Find the false-alarm and hit rates from the correct (unequal-variance) representation and determine what d' the researcher would be expected to calculate using the incorrect (equal-variance) equations.

c. Suppose that the researcher obtains estimates $\widehat{d}'_A$, $\widehat{d}'_B$, and $\widehat{d}'_{AB}$ that equal the values calculated in the previous parts. What conclusion would the researcher make about the dependence or independence of the signals?

7.4. Consider the speaker's voice experiment mentioned at the start of this chapter. Suppose that $I > 2$ different speakers are used. During the test phase, the words must be assigned to one of these I alternatives. Would you expect the results to be consonant with a unidimensional model (as in Figure 7.6)? Think of a set of speakers and speculate on the dimensions that might be involved.

Chapter 8

Finite-state models

The detection and discrimination models described thus far in this book assign Gaussian distributions to the latent random variables X_n and X_s that represent the effect of a stimulus. Values of these variables are real numbers. Thus, the evidence for or against a particular alternative that is provided by a stimulus is graded. Although a treatment based on the continuous Gaussian distribution is by far the most popular model in signal-detection theory, it is not the only alternative. An important class of models uses a small number of discrete states to represent the effects of a stimulus. This chapter describes some models of this type.

It is helpful to think of the finite-state models in this chapter somewhat differently from the continuous models. Although it is possible to construct them by placing a criterion on a continuous axis (as in Figure 8.2, below), it is easier to think of them as describing what happens when the stimulus puts the observer into one of a small number of distinct internal states. These states are not directly observable, just as X_s and X_n were not observable, but they provide the information that the observer uses to make a response. Because the number of internal states of the observer is small and countable, they are known as *finite-state models*. These states also can be interpreted as indicating when the strength of a stimulus passes some threshold, so they are also known as *threshold models*.

8.1 The high-threshold model

The simplest and most important of the finite-state models is known as the *high-threshold model*. In it, all the observer's information about the stimulus is expressed by a single binary variable. After any stimulus has been presented, the observer is left in one of two latent states, the *detect state* D or the *uncertain state* U. The observer then uses the current state to make a response.

For a pair of states to carry useful information, they must be differentially associated with the presence or absence of the signal. Moreover, the state cannot accurately mirror the state of the stimulus. If a signal always gave state D and a noise event always gave state U, then an efficient observer could make a correct response simply by reporting the internal state. A more interesting model, and one that is most consistent with the spirit of signal-detection theory, posits a probabilistic relationship between the internal state and the external events. Thus, it incorporates the variability that is found in an observer's responses.

The high-threshold model is based on four assumptions. Two of these assumptions concern the way that the internal response is related to the stimulus:

1. Noise events never lead to the detection state:

$$P(D|\text{noise}) = 0 \qquad \text{and} \qquad P(U|\text{noise}) = 1.$$

2. Signal events lead to the detection state with probability α (the Greek letter alpha):

$$P(D|\text{signal}) = \alpha \qquad \text{and} \qquad P(U|\text{signal}) = 1 - \alpha.$$

The probability α reflects the strength of the signal. It is near zero when the signal is little different from noise and near one when the signal is strong and nearly always detected. Thus, it plays a role in the finite-state theory that is similar to that played by d' or μ_s in the Gaussian models. In the high-threshold model, α replaces the various detectability measures mentioned in Chapter 4.

The second part of the model connects the internal event to the observer's overt response. When the detection state D occurs, the appropriate response is clear. This state never arises from noise, so the observer can always say YES to report a signal. The nondetect state U can arise from either a signal event or a noise event. The model here asserts that following this event the observer guesses, saying YES with probability γ (the Greek letter gamma). The second pair of assumptions express this portion of the model:

3. When state D occurs, the observer reports a signal:

$$P(\text{YES}|D) = 1 \qquad \text{and} \qquad P(\text{NO}|D) = 0.$$

4. When state U occurs, the observer guesses with probability γ:

$$P(\text{YES}|U) = \gamma \qquad \text{and} \qquad P(\text{NO}|U) = 1 - \gamma.$$

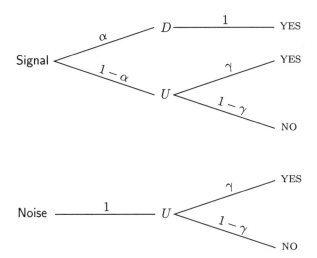

Figure 8.1: The tree diagram used to calculate the conditional response probabilities for the high-threshold model.

The four assumptions of the high-threshold model let one calculate the
probabilities of the four observable outcomes of a trial. A tree diagram
illustrates the analysis. Figure 8.1 shows two trees, one that applies on
signal trials, the other that applies on noise trials. One branch of the signal
tree puts a proportion α of the trials into state D, and the other branch
takes the remaining trials to state U. From state D only YES responses are
made, while from state U, YES and NO responses are made with probabilities
γ and $1 - \gamma$, respectively. In the noise tree, only state U has a positive
probability of occurring, but either YES or NO responses may be made,
based on the same branch associated with that state in the signal tree. The
probability of any path through tree is the product of the probabilities along
its limbs, and the probability of a response is the sum of the probabilities
of all paths that end with that response. In the upper tree (signal stimuli)
there are two paths that end with YES. These have probabilities α and
$\gamma(1-\alpha)$. The hit rate is the sum of these two values:

$$P_H = \alpha + \gamma(1-\alpha). \tag{8.1}$$

In the noise tree, the only branch ending on YES has probability γ, giving
the false-alarm probability:

$$P_F = \gamma. \tag{8.2}$$

A more formal approach to this calculation uses the rules of probability
theory. Suppose that a stimulus A (either signal or noise) is presented and

the observer makes the response R (either YES or NO). The probability of response R given stimulus A is broken up by the internal state using the law of total probability (Equations A.5 and A.10):

$$P(R|A) = P(R \cap D|A) + P(R \cap U|A)$$
$$= P(R|D \cap A)P(D|A) + P(R|U \cap A)P(U|A).$$

A fundamental tenet of all the signal-detection models is that the response depends on the stimulus only through the internal state. As a result, the stimulus can be dropped from the conditional probability of a response when the internal state is given:

$$P(R|D \cap A) = P(R|D) \qquad \text{and} \qquad P(R|U \cap A) = P(R|U).$$

After this simplification, the response probability is

$$P(R|A) = P(R|D)P(D|A) + P(R|U)P(U|A).$$

Applied to the high-threshold model this equation yields Equations 8.1 and 8.2:

$$P_H = P(\text{YES}|D)P(D|\text{signal}) + P(\text{YES}|U)P(U|\text{signal}) = \alpha + \gamma(1-\alpha),$$
$$P_F = P(\text{YES}|D)P(D|\text{noise}) + P(\text{YES}|U)P(U|\text{noise}) = 0 + \gamma.$$

These equations are the finite-state analogs of the integrals for the Gaussian model in Equations 3.1 on page 45.

Equations 8.1 and 8.2 express the response probabilities in terms of the theoretical parameters α and γ. To apply the model, one needs estimates of these parameters based on the observed hit rate h and false-alarm rate f. Because the number of parameters equals the number of independent data values, the problem is easy. Substitute the observed values h and f for the probabilities P_H and P_F, and replace the parameters by their estimates, turning Equations 8.1 and 8.2 into

$$h = \hat{\alpha} + \hat{\gamma}(1-\hat{\alpha}) \qquad \text{and} \qquad f = \hat{\gamma}.$$

Now solve this pair of simultaneous equations to get the estimators:

$$\hat{\alpha} = \frac{h - f}{1 - f} \qquad \text{and} \qquad \hat{\gamma} = f. \tag{8.3}$$

Because the first of these equations uses the false alarm rate to eliminate the effects of guessing from the hit rate, it is sometimes known as the *correction for guessing*.

Example 8.1: In the first session in the experiment described in Section 1.2, the response rates were $h = 0.82$ and $f = 0.46$. Gaussian-model estimates were made in Example 2.2 on page 24. Interpret these results in terms of the high-threshold model.

Solution: Equations 8.3 apply. The estimates are

$$\hat{\alpha} = \frac{h - f}{1 - f} = \frac{0.82 - 0.46}{1 - 0.46} = 0.67,$$

$$\hat{\gamma} = f = 0.46.$$

By the high-threshold model, the observer detected the signal on two thirds of trials on which it was presented. On the noise trials and the other third of the signal trials, the observer guessed, saying YES and NO about equally often (actually, 46% and 54%, respectively). When the trials on which guessing occurred are excluded, the observed hit rate of 82% drops to the estimated detection rate of 67%.

The high-threshold model can also be constructed from a pair of distributions and a criterion. For the distributional model to agree with the calculations above, the random variables X_n and X_s must have two properties. First, there must be a "threshold" above which only the signal distribution has any density. State D arises if and only if the stimulus magnitude exceeds this threshold. The decision criterion lies somewhere below this high threshold. Second, below the threshold, the placement of the decision criterion must affect the probabilities for the noise and signal distributions comparably. The ratio of the heights of the two density functions $f_n(x)$ and $f_s(x)$ here must be unrelated to x. These properties are most easily represented (although not uniquely) by the pair of uniform distributions shown in Figure 8.2. The noise random variable X_n has a uniform distribution between 0 and 1. The threshold is at the top of this distribution, at $x = 1$, so that all of the noise distribution falls below it. The signal random variable X_s has a wider range between 0 and $1/(1-\alpha)$. A uniform distribution over this range has a height of $1 - \alpha$, and a proportion α of it lies above the threshold. This area corresponds to $P(D|\text{signal})$. Decisions are made by placing a criterion at $\lambda = P(\text{NO}|U) = 1 - \gamma$. The area in the lower distribution above this criterion gives the hit rate. It equals the value calculated in Equation 8.1.

The distributional picture of Figure 8.2 fits the high-threshold and Gaussian models into a similar framework and shows the relationship between them. In the Gaussian model, the entire range of values of X are informative, while in the high-threshold model the only information available is whether the observation is above or below the threshold. The fact that all of the noise distribution falls below this threshold gives the model

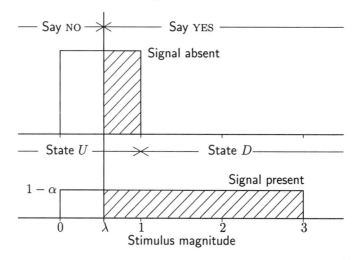

Figure 8.2: *A pair of uniform distributions and a criterion λ that produce the probabilities of the high-threshold model. Areas corresponding to false alarms and to hits are shaded. The parameter values (α = 0.67 and γ = 0.46) match the estimates in Example 8.1. Compare this figure to the Gaussian representation in Figure 1.3 on page 14.*

its name. However, introducing the numerical axis and the distributions adds complexity to the representation that is not discernable in the data. The finite-state interpretation of the high-threshold model is simpler and more in character with the properties of the model.

Both the parameter α of the high-threshold model and the parameter d' of the equal-variance Gaussian model are quantities that describe the detectability of a signal. Both parameters measure roughly the same thing, but they derive from different representations of the detection process. It is important to understand the circumstances under which each model applies. The most salient difference between the models is the extent to which the observer has access to graded information about the stimulus. The high-threshold model allows the observer only two levels of information (the states D and U), while the Gaussian model allows for a continuous range of information (the observation X). If the stimuli are such that the observer can accumulate complex or partial information, then the two-state representation of the high-threshold model is inappropriate. In such cases the "correction for guessing" embodied in Equation 8.3 is not meaningful and the high-threshold model should not be used. If the signal is very simple or the observer has a very limited ability to extract information from it, then a dichotomous model may be more appropriate than a graded one.

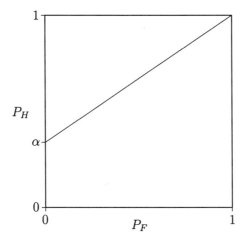

Figure 8.3: The high-threshold operating characteristic with detection probability α.

8.2 The high-threshold operating characteristic

Although one often must argue for or against the high-threshold model based on general characteristics of the detection process, sometimes one can appeal to data. As with the assumptions about the variance discussed in Chapter 3, the operating characteristic is diagnostic. Its shape is different under the two models, and with the right data it can be used to distinguish between them.

The operating characteristic predicted by the high-threshold model is obtained by solving Equations 8.1 and 8.2 to write P_H as a function of P_F. The false-alarm rate is substituted for the bias parameter γ in Equation 8.1 to give

$$P_H = (1-\alpha)P_F + \alpha. \qquad (8.4)$$

This linear equation implies that the operating characteristic is a straight line when plotted in probability coordinates. Figure 8.3 shows this function. The operating characteristic begins at a hit rate of α on the left when $\gamma = 0$ and the observer never guesses YES. As γ increases to one, it moves directly to the point $(1, 1)$ at the corner of the graph, where the observer says YES under every circumstance.

Figure 8.4 compares the operating characteristics of the Gaussian and high-threshold models. Each graph shows two operating characteristics, one

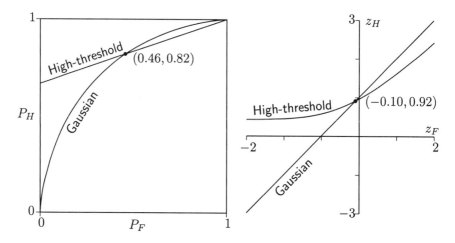

Figure 8.4: *Operating characteristics for the high-threshold and Gaussian models, plotted on probability and Gaussian axes.*

from the high-threshold model and one from the equal-variance Gaussian model. Each line passes through the point obtained in Example 8.1. In the left panel, the axes of the plot are the probabilities of false alarms and hits, and the common point is $(P_F, P_H) = (0.46, 0.82)$. In the right-hand panel, the axes have been rescaled to Gaussian coordinates and the common point $[Z(P_F), Z(P_H)]$ is $(-0.00, 0.92)$. Each model is simplest in its own domain: the high-threshold model gives a straight line in the linear domain, and the Gaussian model gives a straight line in the transformed domain.

In principle, a plot of the operating characteristic lets one check which of the models applies to a set of data. One observation of an (f, h) pair cannot discriminate between the models. As Figure 8.4 illustrates, an operating characteristic of either type can be passed through any single point. At a minimum, one needs data from two detection conditions, produced either by varying the bias or by using confidence ratings. The data are plotted both in the basic probability space and in Gaussian coordinates. If the high-threshold model is correct, then the two points lie on a straight line that passes through $(1, 1)$ in the direct plot (except for sampling error). If the equal-variance Gaussian model is correct, then the transformed points lie on a straight line with unit slope.

Discrimination between the two models is easier when three or more points are available. These allow one to investigate whether the operating characteristic is a straight line in either of the spaces, without regard to ancillary properties such as passage through $(1, 1)$ or unit slope. Either of these properties can fail in generalizations of the high-threshold or

F.A.	f	$Z(f)$	Hits	h	$Z(h)$
42	0.168	−0.962	145	0.580	0.202
81	0.324	−0.456	172	0.688	0.490
112	0.448	−0.130	188	0.752	0.680
174	0.696	0.512	214	0.856	1.062

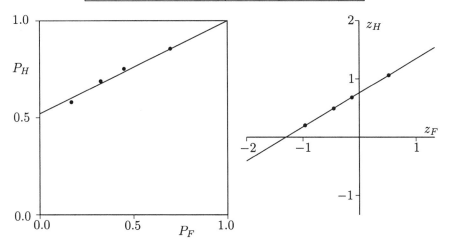

Figure 8.5: *Detection conditions plotted in probability and Gaussian coordinates.*

equal-variance Gaussian models without compromising their finite-state or Gaussian character. The linearity of the operating characteristic is more fundamental. Of course, it is possible that the points do not fall on a straight line in either space, in which case neither a high-threshold nor a Gaussian model is correct.

A more common occurrence than the failure of both models is to find that both can be fitted reasonably well. When the points are not widely separated, both models give similar predictions. The following example illustrates the difficulty.

Example 8.2: In four 500-trial detection conditions (250 signal and 250 noise stimuli), the numbers of false alarms and hits are (42, 145), (81, 172), (112, 188), and (174, 214). Show that these data can be fitted by the high-threshold model. How satisfactory is the Gaussian model?

Solution: The data are converted to proportions and plotted in the left-hand part of Figure 8.5. A straight line from the upper-right corner of the box passes close to the four points, indicating that the high-threshold model can describe the data—the slight deviations from the line can easy be due to sampling fluctuation. The fitted line strikes the ordinate at about 0.52, which is the estimate of α.

The right-hand plot in Figure 8.5 shows the points in Gaussian coordinates. These four points also fall nearly on a straight line, indicating that they could be fitted by a Gaussian model. The line through the points does not lie at a 45° angle, so an unequal-variance model is needed. The parameter estimates for the fitted model are $\hat{\mu}_s = 1.30$ and $\hat{\sigma}_s = 1.71$, with an operating-characteristic area of $\widehat{A_z} = 0.74$. Because the unequal-variance model is needed, the Gaussian representation is more complicated than the high-threshold representation, requiring three parameters (μ_s, σ_s^2, and λ) instead of two (α and γ).

8.3   Other finite-state representations

This section describes several other versions of the finite-state detection model. Although these are not as widely used as the high-threshold model, they sometimes characterize a detection situation better than it does.

An important assumption of the high-threshold model is that the detect state is never produced by a noise stimulus—it is this property that gives it its high-threshold characteristic. One way to modify the model while keeping its two-state nature is to reverse this assumption. In the *low-threshold model*, the signal always leaves the observer in the same state, and a noise stimulus leaves the observer in either of two states. When this model is given a distributional interpretation, as in Figure 8.2, the threshold is placed below the smallest of the signal-induced events, giving the model its name.

As in the high-threshold model, let U be an uncertain state that can arise from either a signal or a noise event. Instead of the detect state of the high-threshold model, the low-threshold model has a *nondetect state* N that arises only from the noise event. Denote that probability by β (which here is not the likelihood ratio). The conditional probabilities of the two states given a stimulus are

$$P(U|\text{noise}) = 1 - \beta, \qquad P(N|\text{noise}) = \beta,$$
$$P(U|\text{signal}) = 1, \qquad P(N|\text{signal}) = 0.$$

A plausible strategy for the observer is to respond NO when state N occurs and to guess in state U, leading to the response rules

$$P(\text{YES}|U) = \gamma, \qquad P(\text{NO}|U) = 1 - \gamma,$$
$$P(\text{YES}|N) = 0, \qquad P(\text{NO}|N) = 1.$$

Derivations of the hit rate and false-alarm rate for this model exactly parallel those of the high-threshold model (see Problem 8.3). The operating characteristic for the low-threshold model is a straight line connecting the point $(0, 0)$ to the point $(1 - \beta, 1)$.

The circumstances under which one may decide to apply the low-threshold model to a set of data are similar to those that suggest the high-threshold model. The situation must be one in which the evidence available from the signal is both impoverished and all-or-none in character. The distinction between the threshold models lies in which stimulus type leads to an uncertain state. If the signal can sometimes be overlooked, then the high-threshold model is the better representation, while if the noise event can be confused with a signal, then the low-threshold model should be used. A more complete investigation of the models (and of the Gaussian model) requires enough data to plot an operating characteristic, as was done in Figure 8.5.

The high-threshold and low-threshold models can be combined in a *three-state model* that allows both certainty and ambiguity to follow either signal or noise events. It combines the sure-detect properties of the low-threshold model with the sure-exclude properties of the high-threshold model. The three states are a detect state D, a nondetect state N, and an uncertain state U. The relationship of these states to the stimulus and response is shown in the tree diagram of Figure 8.6. State D occurs only in the presence of the signal, and state N occurs only in the presence of noise. Let α and β be the conditional probabilities of these events:

$$P(D|\text{signal}) = \alpha, \qquad P(U|\text{signal}) = 1 - \alpha, \qquad P(N|\text{signal}) = 0. \quad (8.5)$$

$$P(D|\text{noise}) = 0, \qquad P(U|\text{noise}) = 1 - \beta, \qquad P(N|\text{noise}) = \beta. \quad (8.6)$$

The responses YES and NO are always made from the states D and N, respectively. As in the high-threshold and low-threshold models, guessing occurs in state U, with the YES response being made with probability γ.

The hit rate and false-alarm rate for the three-state model are found by applying the law of total probability, this time with three events:

$$\begin{aligned} P_H &= P(\text{YES}|D)P(D|\text{signal}) + P(\text{YES}|U)P(U|\text{signal}) \\ &\quad + P(\text{YES}|N)P(N|\text{signal}) \\ &= \alpha + \gamma(1-\alpha) \\ P_F &= P(\text{NO}|D)P(D|\text{signal}) + P(\text{NO}|U)P(U|\text{signal}) \\ &\quad + P(\text{NO}|N)P(N|\text{signal}) \\ &= \gamma(1-\beta) \end{aligned}$$

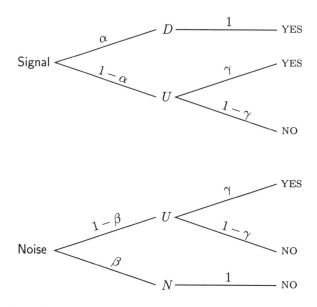

Figure 8.6: A tree diagram used to calculate the conditional response probabilities for the three-state model.

As the guessing probability γ goes from zero to one, the operating characteristic traces out a straight line from the point $(0, \alpha)$ to the point $(1-\beta, 1)$, as shown in Figure 8.7. The model exactly fits any pair of detection conditions that fall on a line that does not cross the major diagonal of the operating characteristic space.

The three-state model generalizes the high-threshold and low-threshold models in the same way that the unequal-variance Gaussian model generalizes its equal-variance version. It contains both simpler models as special cases. When $\beta = 0$, the sure-noise state N never occurs, and the model becomes the two-state high-threshold model. When $\alpha = 0$, the sure-detect state D never occurs, and the model becomes the low-threshold model. This hierarchical structure is useful in investigating the class of threshold models. If enough data are available, the three-state model can be fitted and compared to restricted versions in which $\alpha = 0$ or $\beta = 0$.

Having come this far, it is tempting to extend the picture by including other states associated with intermediate degrees of certainty. For example, a four-state model could have two uncertain states, U_1 and U_2, with different guessing probabilities γ_1 and γ_2. However, such models are impractical. One problem is that the extra states add greatly to the number of parameters. Splitting state U adds three new parameters to the model: two to relate the stimulus to the new state (comparable to α and β) and

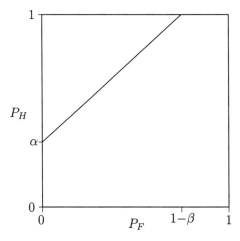

Figure 8.7: *The operating characteristic for the three-state model.*

one to give the new guessing rate. These six parameters cannot be estimated accurately from any reasonable-sized set of data. A second problem with multiple-state models is that they are hard to discriminate from models based on continuous random variables. The ambiguity that appeared in Example 8.2 is exacerbated. The operating characteristic of a multiple-state model is composed of several straight line segments. Unless very accurate data are available from an exceptionally large number of bias conditions, this polygonal function cannot be distinguished from a smooth curve. If one needs to go beyond the simple threshold models discussed in this chapter, one is usually better served by the continuous Gaussian model.

The only practical way to fit the parameters of the multistate model is to tie them together by adding assumptions that connect the states. For example, one could assume that the distribution of the states under the two types of stimulus has a particular probability distribution, such as a binomial or Poisson distribution. A decision rule based on overlapping distributions of these types is not difficult to construct (an example is given in Section 9.1). Such a model may be sensible when these distributions arise naturally from a theoretical model for the task that is being considered, but again is likely to be very similar to a Gaussian model.

8.4 Rating-scale data

Although the threshold models can be fitted to data from a rating-scale experiment, their all-or-none nature makes them somewhat unnatural. Unlike the Gaussian model, a two-state model has no way to generate distributions of confidence levels that are differentially sensitive to the stimuli. Differences among the confidence levels reflect only guessing. The observer's limited information about the stimulus shows through very clearly in the rating responses.

It is most efficient to analyze the three-state model with r confidence levels first, treating the high-threshold and low-threshold models as special cases of it. Figure 8.8 shows the tree diagram for $r = 4$. No changes are needed in the portion of the model that connects the states N, U, and D to the stimuli (Equations 8.5 and 8.6). Because the detect state D can arise only from the signal, it behooves the observer to make a high-confidence YES response when it occurs. Similarly, the nondetect state N should lead to a high-confidence NO response. The uncertain state U can arise from either noise or signal. When it occurs, no further information is available to help make a confidence rating, so the observer simply distributes the responses randomly among the r confidence levels according to some probability distribution. Denote these probabilities by γ_1, γ_2, ... , γ_r. Collecting the branches of the tree in Figure 8.8 that end on a particular response gives the probabilities shown in the figure. In general, these response probabilities are

$$P(R{=}j|\text{noise}) = \begin{cases} (1{-}\beta)\gamma_j, & j = 1, \ldots, r{-}1, \\ \beta + (1{-}\beta)\gamma_r, & j = r, \end{cases}$$

$$P(R{=}j|\text{signal}) = \begin{cases} \alpha + (1{-}\alpha)\gamma_1, & j = 1, \\ (1{-}\alpha)\gamma_j, & j = 2, \ldots r. \end{cases}$$

Data from a rating experiment consist of two sets of frequencies for the various confidence levels, x_{n1}, x_{n2}, ... , x_{nr} for noise trials and x_{s1}, x_{s2}, ... , x_{sr} for signal trials. The model's parameters can be estimated directly. Responses with intermediate confidence levels arise only from state U. For these responses, the probabilities under signal and noise always stand in the same ratio ρ:

$$\rho = \frac{P(R = j|\text{signal})}{P(R = j|\text{noise})} = \frac{1 - \alpha}{1 - \beta}. \tag{8.7}$$

Responses to the intermediate categories only come from state U. Because their state is known, they can be used to estimate ρ:

$$\widehat{\rho} = \frac{x_{s2} + \cdots + x_{s,r-1}}{x_{n2} + \cdots + x_{n,r-1}}. \tag{8.8}$$

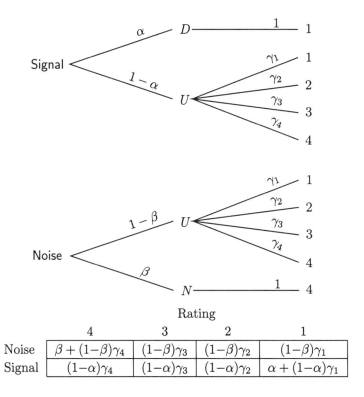

	Rating			
	4	3	2	1
Noise	$\beta+(1-\beta)\gamma_4$	$(1-\beta)\gamma_3$	$(1-\beta)\gamma_2$	$(1-\beta)\gamma_1$
Signal	$(1-\alpha)\gamma_4$	$(1-\alpha)\gamma_3$	$(1-\alpha)\gamma_2$	$\alpha+(1-\alpha)\gamma_1$

Figure 8.8: *A tree diagram for the three-state model with $r = 4$ confidence levels. The table gives the resulting response probabilities.*

Now consider the high-confidence YES responses. Some of these are made from state S, but others may be from state U; the question is, how many. Any such responses on noise trials must be from state U. There are x_{n1} of these responses, so using the ratio $\widehat{\rho}$, about $x_{n1}\widehat{\rho}$ of the high-confidence YES responses on signal trials should also be from state U. The difference, $x_{s1} - x_{n1}\widehat{\rho}$, estimates the number of these responses that were made from state S. Expressing this quantity as a ratio gives an estimate of the probability of entering state S:

$$\widehat{\alpha} = \frac{x_{s1} - x_{n1}\widehat{\rho}}{N_s}. \tag{8.9}$$

A similar argument applied to the other end of the rating distribution is used to estimate β. An estimate of the number of high-confidence NO responses made while in state U is $x_{sr}/\widehat{\rho}$, the number of responses from

state N is obtained by subtracting this value from x_{nr}, and the probability
of entering state N is estimated by the ratio

$$\widehat{\beta} = \frac{x_{nr} - x_{sr}/\widehat{\rho}}{N_n}. \tag{8.10}$$

Either Equation 8.9 or 8.10 can give a negative number. This value implies that the best choice for the probability is zero and that the associated state, D or N, is not required. Then a three-state model is unnecessary, and the data can be fitted by the high-threshold or low-threshold model.

The two-state models are special cases of the three-state model, and their parameters are estimated in almost the same way. The only differences are that one more category is included in the estimate of ρ in Equation 8.8 and that only the relevant one of Equations 8.9 and 8.10 is used. For example, under the high-threshold model the responses with rating r are guesses. Their frequency is added to the numerator and denominator of Equation 8.8, and Equation 8.9 is used to find $\widehat{\alpha}$. For the low-threshold model, the responses with r of 1 are guesses and $\widehat{\beta}$ is calculated from Equation 8.10.

Example 8.3: In a study in which a YES or NO response was followed by a SURE/UNSURE confidence rating, 300 trials of data were collected:

	NS	NU	YU	YS	
Noise	53	38	31	28	150
Signal	19	9	14	108	150

Fit the three-state rating model to these responses. If appropriate, also fit the high-threshold and the low-threshold models.

Solution: Under the three-state model, the central categories NU and YU are only made from state U. The estimate of ρ based on these observations (Equation 8.8) is

$$\widehat{\rho} = \frac{x_{s,NU} + x_{s,YU}}{x_{n,NU} + x_{n,YU}} = \frac{9 + 14}{38 + 31} = 0.333.$$

Using this ratio, Equations 8.9 and 8.10 give estimates of the detection parameters:

$$\widehat{\alpha} = \frac{108 - 28 \times 0.333}{150} = 0.658 \quad \text{and} \quad \widehat{\beta} = \frac{53 - 19/0.333}{150} = -0.027.$$

The negative value for the estimate of β implies that, as far as can be told, state N was not used, and the high-threshold model fits as well as the three-state model. To fit that model, the low-confidence responses are added to the estimate of ρ, giving

$$\widehat{\rho} = \frac{x_{s,\text{NS}} + x_{s,\text{NU}} + x_{s,\text{YU}}}{x_{n,\text{NS}} + x_{n,\text{NU}} + x_{n,\text{SU}}} = \frac{19 + 9 + 14}{53 + 38 + 31} = 0.344.$$

This value only trivially changes the estimate of the detection parameter:

$$\widehat{\alpha} = \frac{108 - 28 \times 0.344}{150} = 0.656.$$

The value of α is clearly nonzero, so the low-threshold model is inappropriate. Plotting the operating characteristic shows as much.

With three or more levels of confidence, the appropriateness of the high-threshold or low-threshold models can be tested, and with four or more levels, the three-state model can be tested. These tests are based on the fact that the intermediate-confidence responses are pure guesses from state U. The distribution of responses to noise stimuli in these categories is proportional to the distribution of responses to signals (Equation 8.7). This property is tested with the chi-square test for association (page 245), using the columns of the data that correspond to guesses. A large value of the test statistic implies that there is a systematic difference between the distribution of ratings for the two types of stimulus, a characteristic that is inconsistent with the finite-state models. When the proportions are not homogeneous, these models should be rejected in favor of a model with graded information.

Example 8.4: Test the rating-scale data in Table 5.1 on page 84 to see if they can be fitted by one of the finite-state models.

Solution: Eliminating the highest-confidence responses (both YES and NO) from the data leaves a 2×4 table containing the intermediate responses:

	N2	N1	Y1	Y2
Noise	161	138	128	63
Signal	65	66	92	136

It is obvious that the distributions in the two rows are not the same—there are more N2 responses for noise stimuli and more Y2 responses for signal stimuli. This result is easily verified statistically. A test of independence in this table with the Pearson statistic (Equation A.54) gives $X^2 = 80.56$ on three degrees of freedom (Equation A.57). A comparison to a chi-square distribution easily rejects independence (the 5% critical value is 7.81 and the 0.1% critical value is 16.27). A continuous-range model such as the Gaussian model is necessary.

Reference notes

A treatment of signal-detection theory from a finite-state perspective is given in Chapter 5 of Atkinson et al. (1965).

Exercises

8.1. Fit the high-threshold model to the data of Problem 2.6.

8.2. An observer in a detection task has a hit rate of 80% with a false-alarm rate of 31%. By rewarding hits and punishing false alarms, the observer is induced to change strategy so that the false-alarm rate falls to 7%. What should the hit rate be under the high-threshold model and the equal-variance Gaussian model?

8.3. Draw a tree diagram comparable to Figure 8.1 for the low-threshold model. Derive the probability of a hit and a false alarm, and show that the operating characteristic is a straight line from the origin to $(1 - \beta, 1)$. Estimate the parameters for the data used in Problem 8.1.

8.4. Modify Figure 8.2 to show the distributions implied by the three-state model of Section 8.3.

8.5. Apply the high-threshold model to the two-alternative forced-choice task of Chapter 6. On each trial, the observer is presented with one signal and one noise stimulus. Each stimulus leads to one of the states D or U. When one stimulus gives the state D, it must be the signal, the observer chooses it, and the response is correct. When both stimuli give state U, the observer chooses randomly between them and is correct with probability $1/2$.

a. Draw a tree diagram for this process, and determine the correct-response probability P_C.

b. Find the area under the high-threshold operating characteristic in Figure 8.3. Note that this value is the same as P_C, as required by the area theorem (Equation 6.15).

8.6. Consider the results from two observers using five-level confidence ratings in a detection task:

	Observer 1					Observer 2				
	5	4	3	2	1	5	4	3	2	1
Noise	16	26	74	82	52	14	50	54	61	71
Signal	107	47	36	46	14	106	46	44	43	11

a. Plot the data from these observers and decide whether either one shows finite-state properties.

b. Fit the three-state model to the data from the better-fitting observer.

c. If either the high-threshold or low-threshold model is appropriate for this observer, fit it.

8.7. The behavior of an ideal observer who uses the high-threshold model is somewhat unexpected. Suppose that the proportion of signals in a series of trials is p_s and that the observer has some knowledge of this value and of α. Use an expanded tree diagram based on Figure 8.1 or the law of total probability to calculate the probability P_C of a correct response. Manipulate this result to show that

$$P_C = 1 - (1-\alpha)p_s + [(1-\alpha)p_s - (1-p_s)]\gamma.$$

Deduce from this expression that the ideal observer will either set γ to zero or to one, depending on the sign of the term in brackets. Describe in words what each of these conditions means. Note: although P_C is maximized by always making the same response from state U, this behavior is not characteristic of many human observers, who are more likely to use a value of γ that is closer to the inferred proportion of signals.

Chapter 9

Likelihoods and likelihood ratios

The decision in the signal-detection models is made by placing a criterion on the axis of an underlying random variable. This procedure is straightforward and intuitive, and it works well with simple symmetrical distributions like the Gaussian. Its use does not require one to be specific about the axis on which the decision is made. Some measure of instantaneous signal intensity suffices, or, more generally, the evidence for (or against) a particular alternative. This axis, and the whole process of probabilistic decision making, is put on firmer mathematical ground by formulating it using a decision strategy known as likelihood-ratio testing. This procedure is equivalent to a criterion cut when the equal-variance Gaussian model holds, but is much more general. It works when the simpler representation does not apply. This chapter introduces likelihood-ratio testing as a description of an observer's decision procedure. Going beyond signal-detection theory, the likelihood methods make contact with a large body of mathematical work in testing under uncertainty and provide a statistical mechanism, mentioned briefly in Section 11.6, to test hypotheses about the observer.

A study of likelihoods and the likelihood-ratio procedure has several things to offer a signal-detection analysis. Most important, it gives a general and principled basis for the decision procedure. It accommodates more complex distributions of evidence than the Gaussian random variables—both those with unusual univariate distributions and those that arise from multidimensional stimuli. In this way, it suggests what the observer may be doing in making a judgment based on a relatively ill-defined attribute such as the "familiarity" of a face. Another important characteristic of the likelihood-ratio procedure is that it makes optimal use of information. It describes how an ideal observer treats a multifaceted source of information.

The discussion in this chapter and the following ones is couched in terms of detection tasks. However, as Chapter 7 indicated, the models for discrimination and identification have the same form as those for detection. With appropriate changes in terminology, this chapter applies to those tasks.

9.1 Likelihood-ratio tests

A brief review of the concepts of statistical inference and statistical decisions sets the context. Consider a statistician who is in the usual business of inferring the characteristics of a population of individuals using a sample drawn from that population. To give a formal structure in which to make this inference, the statistician adopts a probabilistic representation of the situation. The problem is reformulated as an attempt to find out something about a random variable X that represents the quantity or characteristic to be studied. The values that this variable can take constitute the *population* of potential observations.

The next step is to assume that certain things are known about the random variable. For example, X might be asserted to be continuous, to have a finite range or a Gaussian distribution, and so on. Certain other characteristics are unknown and are the target of the investigation. When these characteristics are numbers, they are referred to as *parameters* of the population. For example, in the commonly used t test, the population is assumed to be Gaussian with unknown mean and variance, and the mean is the target. To find out the values of these parameters, the statistician treats the sample of data as a set of instances of the random variable. These observations give evidence about the population parameters, but that information is imperfect. Because the observations are of a random variable, their properties are not exactly the same as those of the population. Hypotheses about the population parameters must be tested with information extracted from the sample, but the imperfect correspondence between the population and sample inevitably leaves a chance of error.

The statistician may be interested in solving either of two somewhat different problems. In *estimation* the sample is used to make a best guess at the value of an unknown parameter. For example, the statistician may approximate the unknown mean of a population by the average of a sample of 50 scores. In *hypothesis testing*, the goals are more limited. The statistician has two or more hypotheses about the population random variable and, on the basis of the sample, wants to select one of these as the most satisfactory.

Many different procedures have been developed for hypothesis testing. In much scientific research, null-hypothesis testing is common, in which a specific null hypothesis (there is no effect) is contrasted with a general alternative hypothesis (there is some effect). The asymmetry of these hypotheses introduces corresponding asymmetries in the conclusions (one can reject the null hypothesis but not the alternative). Signal-detection theory is closer to another hypothesis testing procedure, known as likelihood-ratio testing. In it, the hypotheses being tested are equally specific, and the test lets one make a decision in favor of one of them. The conclusions from such tests are symmetrical.

The likelihood-ratio procedure gives a basis for both estimation and testing. A simple, although very artificial, example involving a biased coin serves to show how testing works. Suppose that there exist two populations of coins. When any coin in the first population is thrown, the probability that it will come up heads is 0.6, and when any coin in the second population is thrown, this probability is 0.7. The coins are otherwise indistinguishable. An observer is given a coin and must decide to which population it belongs. To make the decision, the observer collects data by tossing the coin. One throw is not much use—the two populations are similar enough that observing one heads or one tails tells little. So the coin is flipped several times and the number of heads is recorded. A relatively large number of heads favors 0.7, and a smaller number favors 0.6.

More formally, the outcome of the coin toss is treated as a random Bernoulli event. Denote the probability that a head occurs by π (here a probability, not $3.14159\ldots$). The statistical observer has two hypotheses about this population parameter:

$$H_1: \pi = 0.6,$$
$$H_2: \pi = 0.7.$$

A sample of N observations of the random event is collected by throwing the coin N times. The observer must now decide how to use the result of these throws to select one of the hypotheses.

Specifically, suppose that $N = 32$ tosses are made and heads occurs $x = 20$ times. Although the observed value of x depends on the value of π, the observed proportion x/N is not the true value of π. This ratio is 0.625, which does not equal either of the hypothesized values, nor can any other proportion based on 32 observations do so. However, the observed value of x is reasonably consistent with either hypothesis.

The key to the making a choice between the hypotheses is that the observer knows quite a bit about the distribution of x. It is easy to calculate the probability that $x = 20$ under each hypothesis. In a sequence of N independent and identical random Bernoulli events, the number of times x that one of the alternatives is obtained has a binomial distribution (Equation A.35 on page 236):

$$P(x; N, \pi) = \frac{N!}{x!(N-x)!} \pi^x (1-\pi)^{N-x}. \tag{9.1}$$

If the hypothesis H_1 that $\pi = 0.6$ were true, then the probability of the observation $x = 20$ would be

$$P(x = 20; H_1) = \frac{32!}{20!\,12!}(0.6)^{20}(0.4)^{12} = 0.1385.$$

If the hypothesis H_2 were true, then this probability would be somewhat smaller:

$$P(x = 20; H_2) = \frac{32!}{20!\,12!}(0.7)^{20}(0.3)^{12} = 0.0958.$$

A natural way to decide between the two hypothesis is to choose the one that makes the data the most likely to have been observed. Because $P(x|H_1)$ is greater than $P(x|H_2)$, the first hypothesis is chosen, and one decides in favor of $\pi = 0.6$. The likelihood-ratio decision rule formalizes this strategy.

The use of the binomial distribution to make this decision differs from its use in an ordinary probability calculation. Normally, one uses the distribution function to find the probability of data (here the value of x) from known values of the parameters of the distribution (here the quantities N and π). Equation 9.1 gives this probability. In contrast, the observer here knows the data, but wants to find out about the parameters. Equation 9.1 is still used to aid in the decision, but it can no longer be interpreted as a probability. The values of N and x are determined by the coin-tossing experiment, so are fixed. Only π remains free. This parameter is not the realization of a random variable (as x was), so what is calculated is not a probability, least of all the probability that one of the hypotheses is correct. To emphasize this distinction, when a probability function is used in this way, it is known as a *likelihood* and is given a different symbol:

$$\mathcal{L}(\pi) = \frac{N!}{x!(N-x)!} \pi^x (1-\pi)^{N-x}. \tag{9.2}$$

The right-hand sides of Equations 9.1 and 9.2 are identical, but they are conceptualized differently, as implied by the different left-hand sides. The left-hand side of Equation 9.1 indicates that it gives the probability of the event x, while that of Equation 9.2 indicates that it gives a measure that depends on π.

An observer who must choose between the hypotheses H_1 and H_2 does so by calculating the likelihoods, $\mathcal{L}(H_1)$ and $\mathcal{L}(H_2)$, associated with these hypotheses. The hypothesis with the larger likelihood is chosen by applying the *decision rule*

$$\begin{cases} \text{If } \mathcal{L}(H_1) > \mathcal{L}(H_2), \text{ then select } H_1, \\ \text{If } \mathcal{L}(H_1) < \mathcal{L}(H_2), \text{ then select } H_2. \end{cases}$$

When the two likelihoods are exactly equal, they provide no evidence that favors one hypothesis over the other. In such cases, the observer could equivocate or return to seek more data. However, in many applications the probability of an exact tie is zero, or so close to it, that it may be assigned arbitrarily to either decision without any practical effect. In this book, the decision rules are written to assign ties to the first alternative:

$$\begin{cases} \text{If } \mathcal{L}(H_1) \geq \mathcal{L}(H_2), \text{ then select } H_1, \\ \text{If } \mathcal{L}(H_1) < \mathcal{L}(H_2), \text{ then select } H_2. \end{cases} \tag{9.3}$$

Applying this decision rule lets one select one of the two hypotheses based on the data, exactly as was done less formally above.

Instead of writing the decision rule as a direct comparison of $\mathcal{L}(H_1)$ to $\mathcal{L}(H_2)$, it is more convenient to convert them to a single number before making the decision. The *likelihood ratio* for two hypotheses is the quotient

$$\mathcal{R}(H_1 : H_2) = \frac{\mathcal{L}(H_1)}{\mathcal{L}(H_2)}. \tag{9.4}$$

A decision to accept one hypothesis or the other is made by comparing this ratio to a criterion C:

$$\begin{cases} \text{If } \mathcal{R}(H_1 : H_2) \geq C, \text{ then select } H_1, \\ \text{If } \mathcal{R}(H_1 : H_2) < C, \text{ then select } H_2. \end{cases} \tag{9.5}$$

If there is no reason to prefer one hypothesis over the other, as in Rule 9.3, then $C = 1$. Otherwise, the value of C can be shifted to favor one of the hypotheses.

The likelihood ratio needed to choose between the hypothesis H_1 that $\pi = \pi_1$ and the hypothesis H_2 that $\pi = \pi_2$ is found by substituting Equation 9.2 into the likelihood ratio:

$$\mathcal{R}(H_1 : H_2) = \frac{\mathcal{L}(H_1)}{\mathcal{L}(H_2)}$$

$$= \frac{\dfrac{N!}{x!(N-x)!}\pi_1^x(1-\pi_1)^{N-x}}{\dfrac{N!}{x!(N-x)!}\pi_2^x(1-\pi_2)^{N-x}}$$

$$= \left(\frac{\pi_1}{\pi_2}\right)^x \left(\frac{1-\pi_1}{1-\pi_2}\right)^{N-x}. \tag{9.6}$$

Decisions are based on this quantity.

Example 9.1: Use the likelihood ratio to choose between the hypotheses $\pi = 0.6$ and $\pi = 0.7$ based on $x = 20$ successes in $N = 32$ trials.

Solution: The likelihood calculation using Equation 9.6 is

$$\mathcal{R}(0.6 : 0.7) = \left(\frac{0.6}{0.7}\right)^{20} \left(\frac{0.4}{0.3}\right)^{12} = 1.45.$$

This ratio exceeds the criterion $C = 1$ for an unbiased decision, so the hypothesis $\pi = 0.6$ is favored.

Numerically the likelihood ratio has two inconvenient characteristics. First, it can be a very small or a very large number. For example, had the coin been flipped 1000 times with 575 heads, one would find that $\mathcal{L}(\pi=0.6) = 0.006997$, $\mathcal{L}(\pi=0.7) = 1.739 \times 10^{-17}$, and the likelihood ratio is 4.023×10^{14}. Although such very large values (or the corresponding very tiny values) do not hamper the decision—these results strongly favor $\pi = 0.6$—they can lead to computational instability and rounding errors. Second, the likelihood ratio is asymmetrical. Likelihood ratios of 3 and $\frac{1}{3}$ represent support of the same strength for H_1 and H_2, respectively, although they lie at different arithmetic distances from the neutral value of 1. Both these inconveniences are avoided by working with the (natural) logarithm of the likelihood ratio instead of the ratio directly. The large likelihood ratio above now is 33.6 and the ratios 3 and $\frac{1}{3}$ correspond symmetrically to -1.10 and $+1.10$.

The change from using a likelihood ratio to using its logarithm has no effect on the decision process. The logarithm is a strictly monotonic function[1] of its argument. Thus, a decision formulated as the likelihood-ratio Rule 9.5 is equivalent to one formulated as the more convenient rule

$$\begin{cases} \text{If } \log[\mathcal{R}(H_1 : H_2)] \geq \lambda, \text{ then select } H_1, \\ \text{If } \log[\mathcal{R}(H_1 : H_2)] < \lambda, \text{ then select } H_2, \end{cases} \tag{9.7}$$

where $\lambda = \log C$. An unbiased decision based on a likelihood-ratio criterion of $C = 1$ becomes a decision based on $\log \mathcal{R}$ with a criterion of $\lambda = 0$. A secondary advantage of the log likelihood ratio is that it has an algebraically

---

[1]A *monotonic function* is one that goes only up or only down, without changes in direction. For a *strictly monotonic* increasing function $f(x)$, if $x > y$, then $f(x) > f(y)$. Without the strictness, $f(x) = f(y)$ is allowed.

simpler form than the likelihood ratio for many distributions, including the Gaussian distribution, the multivariate distributions of Chapter 10, and the sampling distributions used in statistics.

For a decision between two binomial hypotheses, the log likelihood ratio is the logarithm of Equation 9.6:

$$\log[\mathcal{R}(\pi_1 : \pi_2)] = x \log \frac{\pi_1}{\pi_2} + (N - x) \log \frac{1 - \pi_1}{1 - \pi_2}.$$

Applied to Example 9.1, this equation gives

$$\log[\mathcal{R}(0.6 : 0.7)] = 20 \log \frac{0.6}{0.7} + 12 \log \frac{0.4}{0.3} = 0.369.$$

This value is greater than zero, so hypothesis H_1 is chosen. The result is, of course, exactly equivalent to that of Example 9.1, which used the likelihood ratio directly, but it does not require large powers to be calculated.

A random variable with a binomial distribution is discrete. Likelihood-ratio tests can equally well be run with continuous random variables. The distributional information, comparable to the probability function in the discrete case, is given by the probability density function. Instead of probability functions for each hypothesis, $P(X = x; H_1)$ and $P(X = x; H_2)$, one has the density functions $f_1(x)$ and $f_2(x)$. When interpreted as functions of the hypotheses (or the parameters) given the data, these density functions are the likelihoods $\mathcal{L}(H_1)$ and $\mathcal{L}(H_2)$. The log likelihood ratio compares them:

$$\log \mathcal{R}(H_1 : H_2) = \log \frac{\mathcal{L}(H_1)}{\mathcal{L}(H_2)} = \log f_1(x) - \log f_2(x). \tag{9.8}$$

A decision is made by applying a criterion, as in Rule 9.7.

The log likelihood ratio has a different distribution depending on which of the two hypotheses is actually true. When H_1 holds, the observation x is most likely to come from a region where $f_1(x)$ is larger than $f_2(x)$, and the log likelihood ratio is positive. When H_2 is true, the reverse holds. Figure 9.1 shows these distributions, with the criterion λ that separates them in Rule 9.7. The zero point on the log likelihood-ratio axis lies where the two distributions cross. Except for this change of origin, this picture is essentially identical to the one adopted by signal-detection theory. For the symmetrical distributions shown in Figure 9.1, the shift of origin gives the criterion the same value as the centered criterion λ_{center} shown in Figure 2.3 on page 27.

Example 9.2: A random variable X has a Gaussian distribution with variance $\sigma^2 = 1.0$. The value of its mean is known to be either 1.8 or 2.3.

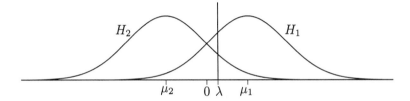

Figure 9.1: *The distributions of the log likelihood ratio when each of the hypotheses H_1 and H_2 is correct. The criterion λ gives the decision point.*

The random variable is observed to take the value $x = 2.1$. Choose one of the putative means based on a likelihood-ratio test.

Solution: Let H_1 and H_2 correspond to the means of 1.8 and 2.3, respectively. The likelihoods, calculated from the density function in Equation A.42 on page 237, are

$$\mathcal{L}(H_1) = \frac{1}{\sqrt{2\pi}}e^{-(2.1-1.8)^2/2} = 0.3814,$$

$$\mathcal{L}(H_2) = \frac{1}{\sqrt{2\pi}}e^{-(2.1-2.3)^2/2} = 0.3910.$$

The log likelihood ratio is

$$\log \mathcal{R}(H_1 : H_2) = \log \mathcal{L}(H_1) - \log \mathcal{L}(H_2) = -0.25.$$

The value is negative, so an observer who is unbiased, in the sense of maximizing correct responses when the alternatives are equally likely, chooses H_2.

9.2 The Bayesian observer

Although the likelihoods are calculated using probability functions or probability density functions, they are not probabilities. An observation x is a probabilistic quantity in the original distribution, but when one fixes it, there is no longer any random variable that has a distribution. For example, although the likelihood $\mathcal{L}(H_1)$ in the binomial example above equaled 0.1385, there is no event that has this probability. It is certainly not the probability that H_1 is true. The lack of a probability value for the hypothesis is unfortunate. When deciding between two hypotheses, one would like to choose the hypothesis that has the greatest probability of being correct, not just the one with the greatest likelihood. However, in order to speak about how likely the hypotheses are to be true, some further assumptions about the observer must be made. Adding these assumptions turns a likelihood-ratio observer into a *Bayesian observer*.

The root of the likelihood-ratio observer's problem is that there is no direct way to determine hypothesis probabilities from the data alone. A probabilistic model that describes how observations are generated lets one calculate the probability of those observations based on a hypothesis about that model, but not the probability of the hypothesis based on the observations. Roughly speaking, one can find

$$P(\text{data}|\text{hypothesis}), \quad \text{but not} \quad P(\text{hypothesis}|\text{data}).$$

It is quite possible for a set of data to be consistent with a hypothesis, yet for that hypothesis to be completely implausible on other grounds.

To make an assertion about the probability that a hypothesis is correct, the Bayesian observer makes assumptions about the probabilities of the hypotheses before observing the data. Bayes' Theorem (Equation A.11 on page 228) then can be used to change these probabilities into posterior probabilities that accommodate the data. If the prior probabilities are $P(H_1)$ and $P(H_2)$, and the observer can calculate the likelihoods of the observed event x, $P(x|H_1)$ and $P(x|H_2)$, then the posterior probabilities are

$$P(H_1|x) = \frac{P(x|H_1)P(H_1)}{P(x)} \quad \text{and} \quad P(H_2|x) = \frac{P(x|H_2)P(H_2)}{P(x)}. \quad (9.9)$$

The hypothesis with the largest posterior probability is chosen.

The Bayesian decision maker does not need to calculate the posterior probabilities in full to make the choice. To pick the largest of the two, it suffices to look at the *Bayes' ratio*:

$$\mathcal{B}(H_1 : H_2) = \frac{P(H_1|x)}{P(H_2|x)} = \frac{\dfrac{P(x|H_1)P(H_1)}{P(x)}}{\dfrac{P(x|H_2)P(H_2)}{P(x)}} = \frac{P(x|H_1)P(H_1)}{P(x|H_2)P(H_2)}.$$

The Bayesian observer uses $\mathcal{B}(H_1 : H_2)$ in the same way that the likelihood-ratio observer uses $\mathcal{L}(H_1 : H_2)$. The Bayes' ratio is greater than one when H_1 is more probable than H_2, taking both the prior probabilities and the observation into account, and it is less than one when H_2 is more likely. A Bayes' ratio exactly equal to one gives no support for either hypothesis. The Bayesian observer's decision rule (again assigning equality to the first alternative) is

$$\begin{cases} \text{If } \mathcal{B}(H_1 : H_2) \geq 1, \text{ then select } H_1, \\ \text{If } \mathcal{B}(H_1 : H_2) < 1, \text{ then select } H_2. \end{cases} \quad (9.10)$$

An equivalent decision can be based on any monotonic transform of the Bayes' ratio $\mathcal{B}(H_1 : H_2)$, such as its logarithm. Because the Bayesian observer is working directly with the probabilities that the hypotheses are true, there is no need to introduce an arbitrary criterion into Rule 9.10, as the likelihood-ratio observer did in Rules 9.5 and 9.7.

A Bayesian observer can be interpreted as using a form of likelihood-ratio testing in which the likelihood ratio is adjusted for the prior probabilities. The two probabilities $P(x|H_1)$ and $P(x|H_2)$ are the hypotheses likelihoods $\mathcal{L}(H_1)$ and $\mathcal{L}(H_2)$, so that the Bayes' ratio is equal to the likelihood ratio multiplied by the prior odds:

$$\mathcal{B}(H_1 : H_2) = \frac{\mathcal{L}(H_1)P(H_1)}{\mathcal{L}(H_2)P(H_2)} = \mathcal{R}(H_1 : H_2)\frac{P(H_1)}{P(H_2)}. \qquad (9.11)$$

For an observer who bases the decision on the logarithm of the Bayes' ratio, the adjustment to the log odds ratio is additive:

$$\log[\mathcal{B}(H_1 : H_2)] = \log[\mathcal{R}(H_1 : H_2)] + \text{logit}[P(H_1)]. \qquad (9.12)$$

As this equation shows, the logarithm of the Bayes' ratio equals the log likelihood ratio displaced by the logit of the original hypothesis probability. The same conclusion is reached either by an explicitly Bayesian observer or by a biased likelihood-ratio observer who adopts the criterion

$$\lambda = -\text{logit}[P(H_1)]. \qquad (9.13)$$

Example 9.3: The observer of the coin flips in Example 9.1 knows that twice as many coins with $\pi = 0.7$ were minted as coins with $\pi = 0.6$. How should the observer adjust the decision to accommodate this information?

Solution: The probability that a coin chosen randomly from among all coins minted is of the first type (with $\pi = 0.6$) is $1/3$. If this probability applies to the observer's coin, then the prior probabilities of the two populations are $P(\text{Pop}_1) = 1/3$ and $P(\text{Pop}_2) = 2/3$. The Bayesian observer uses Equation 9.12 and the results of Example 9.1 to calculate

$$\text{logit}[P(\text{Pop}_1)] = \log \frac{1/3}{2/3} = \log 1/2 = -0.693,$$

$$\log[\mathcal{B}(\text{Pop}_1 : \text{Pop}_2)] = \log[\mathcal{R}(\text{Pop}_1 : \text{Pop}_2)] + \text{logit}[P(\text{Pop}_1)]$$
$$= 0.369 + (-0.693) = -0.324.$$

This negative value favors the hypothesis that the coin came from the second population, that in which $\pi = 0.7$.

The equivalent likelihood-ratio observer uses Equation 9.13 to set the criterion in Rule 9.7 to

$$\lambda = -\text{logit}[P(\text{Pop}_1)] = 0.693.$$

The observed likelihood ratio is 0.369, which is less than the new criterion. This observer also classifies the coin as of the second type.

This discussion of the Bayesian observer is written as if x were the observation of a discrete random variable and had positive probability. When x comes from a continuous range, such as when it has a Gaussian distribution, density functions rather than probabilities are used. Although the fact that the probability of any value of x in a continuous distribution is vanishingly small makes calculation of the posterior probabilities by Equations 9.9 problematic, the Bayes' ratio is still well defined:

$$\mathcal{B}(H_1 : H_2) = \frac{f(x|H_1)P(H_1)}{f(x|H_2)P(H_2)} = \frac{f_1(x)}{f_2(x)} \frac{P(H_1)}{P(H_2)}. \qquad (9.14)$$

The same decision rule applies (Rule 9.10), and the same relationship to the likelihood-ratio observer holds (Equations 9.11, 9.12, and 9.13).

Graphically, the placement of a decision criterion according to Equation 9.14 amounts to selecting a criterion where the ratio of the heights of the density functions compensates for the ratio of the prior probabilities. For example, if $P(H_1)$ is thrice as large as $P(H_2)$, then the criterion λ is placed at the point when the density of $f(\lambda|H_1)$ is one third as large as $f(\lambda|H_2)$. The result is to shift the decision point away from the more common density and toward the less common one. The two ratios in Equation 9.14 balance out, leaving a Bayes' ratio of unity at the criterion.

9.3   Likelihoods and signal-detection theory

As it is treated in signal-detection theory, the observer's problem when making a detection or discrimination decision is the same as that of the statistical observer making a likelihood-ratio test. One member of a pair of alternatives (signal or noise) must be chosen, based on uncertain information about which of the pair occurred (the random variable X). What information the observer knows about the probability distribution of the observations is used to formulate the decision strategy. Likelihood-ratio tests, as modified by the Bayesian observer, produce optimal decisions in a wide range of conditions, so it makes sense to assume that the observer— certainly the ideal observer—uses them. This section describes the signal-detection procedure from the likelihood perspective. For the equal-variance

Gaussian model, the development recapitulates results that were presented in earlier chapters, although it gives these results a more substantial foundation. For other models, it gives a new decision rule.

Consider the yes/no detection task and the equal-variance Gaussian model. The two hypotheses to be investigated are that a particular stimulus came on a noise trial and that it came on a signal trial. Denote these hypotheses by N and S, respectively. Information about the stimulus is conveyed by the random variable X, which has the usual distributions under noise and signal,

$$X_n \sim \mathcal{N}(0,1) \quad \text{and} \quad X_s \sim \mathcal{N}(d',1).$$

On a particular trial, an observation x of X is made, and the observer uses it to decide from which distribution it came. The densities of these random variables are the likelihood functions, interpreted as functions of the possible stimulus events N and S, not the observation x. Using the Gaussian density function (Equation A.42), the likelihoods are

$$\mathcal{L}(N) = \varphi(x) = \frac{1}{\sqrt{2\pi}} e^{-x^2/2},$$

$$\mathcal{L}(S) = \varphi(x - d') = \frac{1}{\sqrt{2\pi}} e^{-(x-d')^2/2}.$$

The log likelihood ratio for these hypotheses is

$$\log \mathcal{R}(S : N) = \log \frac{\mathcal{L}(S)}{\mathcal{L}(N)}$$

$$= \log \frac{\frac{1}{\sqrt{2\pi}} e^{-(x-d')^2/2}}{\frac{1}{\sqrt{2\pi}} e^{-x^2/2}}$$

$$= \log[e^{d'(x-d'/2)}]$$

$$= d'(x - d'/2). \tag{9.15}$$

By Rule 9.7, a decision between the alternatives is made by establishing a criterion λ and saying NO or YES depending on whether $\log \mathcal{R}(S : N)$ is below or above this criterion:

$$\begin{cases} \text{If } \log \mathcal{R}(S : N) \leq \lambda, \text{ then say NO,} \\ \text{If } \log \mathcal{R}(S : N) > \lambda, \text{ then say YES.} \end{cases} \tag{9.16}$$

If there is no a priori reason to prefer signals or noise trials, then $\lambda = 0$.

The decision statistic of Equation 9.15 is a linear function of x. As a result, the decision rule based on $\log R$ can be reformulated as a rule based directly on x:

$$\begin{cases} \text{If } x \le \lambda/d' + d'/2, \text{ then say NO,} \\ \text{If } x > \lambda/d' + d'/2, \text{ then say YES.} \end{cases}$$

As this equation shows, with equal-variance Gaussian distributions, the likelihood-ratio procedure leads to the same decision rule that was introduced more intuitively in Chapter 2. The likelihood-ratio formulation gives a stronger theoretical basis for the procedure, particularly by appealing to properties of the distribution of events, rather than to a sensory axis.

By picking the hypothesis that has the greatest chance of being correct given the observations, the Bayesian observer is able to maximize the probability of a correct response. Either the Bayes' ratio is used or the likelihood-ratio criterion λ is adjusted to $-\text{logit}[P(S)]$ (Equation 9.13). This result was discussed in a narrower context in Section 2.5, where it was shown that an observer who maximizes the probability of a correct response should choose a criterion that makes the bias statistic $\log \beta$ equal to minus the logit of the signal probability (Equation 2.15). That observer is now seen to be picking the alternative that has the greatest probability of being correct.

When the noise and signal variances differ in the Gaussian model, the correspondence between a simple criterion and a likelihood-ratio test breaks down. Far out in either tail, the density function of the distribution with the greater variance exceeds that of the distribution with the smaller variance. Figure 9.2 shows a pair of distributions for which the noise variance is 16 times the signal variance. A likelihood-ratio observer will say NO both for small values of X and for very large ones. A single-criterion observer cannot do this. More formally, the densities associated with the two stimuli are

$$f_n(x) = \frac{1}{\sqrt{2\pi}} e^{-x^2/2} \quad \text{and} \quad f_s(x) = \frac{1}{\sqrt{2\pi\sigma_s^2}} \exp\left[-\frac{1}{2}\frac{(x - \mu_s)^2}{\sigma_s^2}\right].$$

Evaluating the likelihood ratio (Equation 9.8) gives

$$\log \mathcal{R}(S : N) = \log f_s(x) - \log f_n(x)$$

$$= \left[\log \frac{1}{\sqrt{2\pi}} - \log \sigma - \frac{(x - \mu_s)^2}{2\sigma_s^2}\right] + \left[\log \frac{1}{\sqrt{2\pi}} + \frac{1}{2}x^2\right]$$

$$= \frac{-1}{2\sigma_s^2}\left[(1 - \sigma_s^2)x^2 - 2x\mu_s + \mu_s^2 + 2\sigma_s^2 \log \sigma_s\right]. \tag{9.17}$$

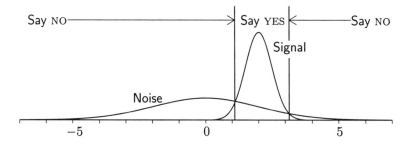

Figure 9.2: *A likelihood-ratio decision rule applied to Gaussian distributions with unequal variances.*

A decision based on this quadratic function—or on the portion within brackets—gives the same result as one based on the two criteria.

Using the likelihood ratio removes the irregularity that was observed in the operating characteristic of a single-criterion observer working with distributions that have different variance. Figure 3.9 on page 52 illustrated this phenomenon when σ_s was twice σ_n. The inability of the single-criterion observer to change the decision when the observation is in the very lower tail causes the operating characteristic (shown also as the dashed line in Figure 9.3) to drop below the diagonal. In contrast, the operating characteristic of the likelihood-ratio observer (Figure 9.3, solid line) is regular and remains above the diagonal. However, unless the difference in variances is extraordinarily large or the centers of the distributions are almost the same, there is little difference between a likelihood-ratio operating characteristic and a criterion operating characteristic (possibly with different parameters). It is impossible to detect the difference between the decision procedures in most realistic experiments. Although the likelihood-ratio model may yield somewhat different estimates of the parameters, in practice it is essentially equivalent to the unequal-variance Gaussian model.

The decision process of the likelihood-ratio observer has some attractive and optimal properties. Among these is the fact that the operating characteristic is always proper; that is, its slope changes from steep to shallow, without reversals, as one moves from the point $(0, 0)$ to the point $(1, 1)$. This form implies that the likelihood-ratio observer makes optimal use of the information in the distributions. The proper form of the function is a consequence of the fact, mentioned in Section 4.5 (page 75), that the likelihood ratio β at the criterion is equal to the slope of the operating characteristic at that point. When the likelihood-ratio observer changes the decision criterion, the slope of the operating characteristic changes with it. The farther a point is from the origin on the operating characteristic, the

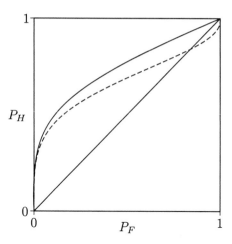

Figure 9.3: Operating characteristics for a single-criterion observer (dashed line) and a likelihood-ratio observer (solid line) based on $X_n \sim \mathcal{N}(0, 1)$ and $X_s \sim \mathcal{N}(1, 4)$. See also Figure 3.9 on page 52.

lower is the criterion and, consequently, the shallower is the slope. So, the likelihood-ratio operating characteristic can take a proper form such as the solid line in Figure 9.3, but never an improper form, such as the dashed line.

As a psychological model, the likelihood-ratio procedure gives a simple description of how decisions are made. From past experience the observer has a feeling for the distribution of effects produced by stimuli from the two conditions. When a new stimulus is presented, the observer refers to these subjective distributions and decides which one was more likely to produce that stimulus. An unbiased observer need not have very exact formal knowledge here—it is only necessary to have enough awareness of the evidence distributions to determine which was greater.

Bias can arise in this scheme in several ways. Inadvertent (and probably inappropriate) biases will occur if the observer incorrectly estimates the likelihoods. Alternatives whose likelihoods are overestimated, perhaps because they are particularly salient, are more often chosen than those that are not. A bias is also introduced if the observer asks that one likelihood be proportionally stronger than the other before selecting it. Such an adjustment can compensate for the perceived costs of the different errors. Most rational is the Bayesian observer, who knows the relative frequencies with which stimuli of the two types occur and uses this information to select the relative strength appropriately.

One advantage of thinking of the decision as the comparison of subjective likelihoods is that it eliminates the need for an sensory continuum, or even for any explicit numerical stimulus axes. A single continuum is natural enough for sensory and psychophysical dimensions such as light or sound intensity, but it is more questionable for other decisions, even sensory ones such as those involving smell or taste. It is still less natural for judgments such as word recognition—the continuum that underlies "familiarity" is difficult to define explicitly—and it is essentially impossible to construct a physical dimension compatible with the medical, seismological, or eyewitness examples in Section 1.1 without appealing to something like the likelihood. The likelihood ratio gives a mechanism by which these complex systems can be reduced to comprehensible process descriptions. Attempts to define a dimension for such tasks often end up by restating the likelihood-ratio principle in other language.

9.4   Non-Gaussian distributions

A likelihood-ratio test does not require the underlying distributions to have any particular form. In particular, they need not be Gaussian. The approach can be used with any distributional form. Although the Gaussian distribution is often a natural choice for a distribution of evidence, another distribution may be more appropriate when it is implied by a theoretical description of the process. For example, in the decision between the two coins discussed in Section 9.1, the binomial distribution is the natural choice for the distribution of outcomes.

Several steps are needed to apply a likelihood-ratio criterion to an arbitrary detection problem. The analysis starts with the noise stimulus N and the signal stimulus S, with their associated random variables X_n and X_s. The distributions of these variables are known by the observer. The first step is to formulate the decision variable. For any observed value x, the log likelihood ratio (Equation 9.8) is

$$\log \mathcal{R}(S : N) = \log \mathcal{L}(S) - \log \mathcal{L}(N) = \log f_s(x) - \log f_n(x). \qquad (9.18)$$

This ratio gives a single number that varies directly with the evidence. The number y obtained by substituting x into Equation 9.18 is used to make the decision. With some distributions, there is a monotonic transformation of $\log \mathcal{R}(S : N)$ that gives a simpler expression but leads to the same decision. The criterion for an optimal decision is zero when the log likelihood is used for y, but can be some other value when a transformation is used. Either way, all the useful information is extracted from the stimulus.

To find the properties of this decision rule, one needs the distributions of the decision variable y under the two stimulus conditions. These random

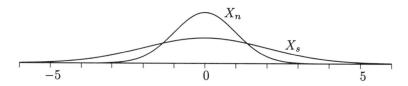

Figure 9.4: *Gaussian distributions for signal and noise events with identical means and different variances. Here* $\mathrm{var}(X_s) = 4\,\mathrm{var}(X_n)$.

variables, Y_n and Y_s, are derived from the original X_n and X_s, but their distributions generally have different forms. Sometimes the form of these distributions can be inferred from their description, as in the binomial case discussed above. Otherwise, their distributions must be derived from those of X_n and X_s. The new random variables are functions of the old ones, given by Equation 9.18:

$$Y_n = \log f_s(X_n) - \log f_n(X_n),$$
$$Y_s = \log f_s(X_s) - \log f_n(X_s).$$

The distributions of Y_n and Y_s are found using the principles of probability theory dealing with the transformation of random variables. The resulting distributions characterize the decision rule. The cumulative distributions give the probabilities of hits and false alarms. The operating characteristic is found by varying the decision criterion λ while holding the distributional aspects of the situation constant.

The likelihood-ratio analysis of Gaussian model in Section 9.3 essentially followed this procedure, although several of the steps were not readily apparent. For the equal-variance model, $\log \mathcal{L}(S : N)$ was a linear function of x (Equation 9.15). As a result, one could take $y = x$, and the original observation appeared to be used directly as the basis of the decision. For the unequal-variance model, x could not be used directly, but decisions based on the quadratic function of Equation 9.17 could be pictured as a dual-criterion decision.

A more obvious illustration of the likelihood-ratio principle comes from applying it to a situation where the distribution of Y is clearly not Gaussian. Consider the problem of distinguishing between two stimuli that differ in their variance. Problems of this general type arise when one is analyzing the power of a time-varying signal, which is a function of its excursion away from its mean. On noise trials, a standard normally distributed quantity is observed, $X_n \sim \mathcal{N}(0, 1)$. On signal trials, the variance of this distribution is increased, so that $X_s \sim \mathcal{N}(0, \sigma^2)$, with $\sigma^2 > 1$. Figure 9.4 shows the density functions for X_n and X_s in the specific case where $\sigma^2 = 4$.

It is immediately clear that a single criterion cannot be used to separate the two distributions. Although the problem could be analyzed with two criteria placed symmetrically on each side of zero, it is simpler to use a derived variable. The first step is to establish a decision variable y. The distributions of X_n and X_s are

$$f_n(x) = \frac{1}{\sqrt{2\pi}} e^{-x^2/2} \quad \text{and} \quad f_x(s) = \frac{1}{\sqrt{2\pi\sigma^2}} e^{-x^2/2\sigma^2}.$$

Calculating the likelihood ratio and simplifying the result gives

$$\log \mathcal{R}(S : N) = \log f_s(x) - \log f_n(x)$$

$$= \left[\log \frac{1}{\sqrt{2\pi}} - \log \sigma - \frac{x^2}{2\sigma^2}\right] - \left[\log \frac{1}{\sqrt{2\pi}} - \frac{x^2}{2}\right]$$

$$= -\log \sigma + \tfrac{1}{2}\left(\frac{\sigma^2 - 1}{\sigma^2}\right)x^2. \tag{9.19}$$

This equation depends on the observation x only through the term x^2 and is monotonically increasing with it. Thus, an equivalent, but simpler, decision variable is the square of the original observation: $y = x^2$.

Now find the distribution of Y for the two stimulus events. The two random variables are $Y_N = X_N^2$ and $Y_S = X_S^2$. Fortunately, no calculation is needed here, as the distribution of the square of a Gaussian random variable is known to be chi-square with one degree of freedom (Equation A.53 on page 245). The random variable Y_n has a standard chi-square distribution, and the random variable Y_s has a chi-square distribution scaled outward by the factor σ^2. The density functions for this particular chi-square distribution can be found by substituting y for x^2 in the normal distribution, giving

$$f_n(y) = \frac{1}{\sqrt{2\pi y}} e^{-y/2} \quad \text{and} \quad f_s(y) = \frac{1}{\sigma\sqrt{2\pi y}} e^{-y/2\sigma^2}. \tag{9.20}$$

These distributions, plotted in Figure 9.5, are very asymmetrical and altogether unlike the Gaussian distributions of conventional signal-detection theory.

The properties of the likelihood-ratio decision are determined from these distributions. The optimal decision point $\lambda^\star$ falls where the densities have the same height: $f_n(\lambda^\star) = f_x(\lambda^\star)$. Equating the two members of Equations 9.20 and solving for $y = \lambda^\star$ gives the optimal criterion:

$$\lambda^\star = \frac{2\sigma^2 \log(\sigma)}{\sigma^2 - 1}.$$

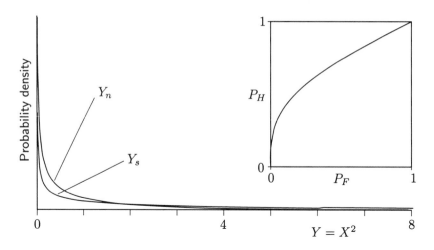

Figure 9.5: *Chi-square distributions of* $Y = X^2$ *with one degree of freedom for the distribution shown in Figure 9.4. The distribution of* Y_s *has considerable mass outside the range plotted here. The inset at upper right shows the operating characteristic.*

For $\sigma^2 = 4$, this criterion lies at $\lambda^\star = 1.85$. This value agrees with the point where the original Gaussian densities in Figure 9.4 cross, which occurs at $\sqrt{1.85} = \pm 1.36$. The cumulative distribution functions $F_n(y)$ and $F_s(y)$ associated with the densities in Equations 9.20 do not have a simple form, but, like their Gaussian equivalents, they can be evaluated numerically. Plotting the points $1 - F_s(\lambda)$ against $1 - F_n(\lambda)$ as λ is varied gives the operating characteristic also shown in Figure 9.5. This operating characteristic is asymmetrical, quite unlike its Gaussian counterpart.

Steps such as these allow one to examine decisions in situations involving complex or irregular sources of information. The approach is particularly useful when working with stimuli that involve several distinct aspects. These multidimensional decisions are discussed in the following chapter.

Reference notes

The likelihood-ratio estimation and testing strategies are covered in many intermediate-level statistical books. A good general treatment, although not always simple, appears in Chapters 18 and 23 of Stewart and Ord (1991) (see also earlier editions), but many other mathematical-oriented statistical texts contain similar information. Bayesian approaches are gaining popularity in statistics, and there are several specialized textbooks de-

voted to them, although these tend to emphasize the statistical and computational aspects of the method, rather than its information-accumulation
function. A discussion of several non-Gaussian detection models is given
by Egan (1975). Johnson, Kotz, and Kemp (1992) and Johnson, Kotz,
and Balakrishnan (1994, 1995) are invaluable compendia of information on
distributions other than the Gaussian. Metz and Pan (1999) discuss the
likelihood-ratio interpretation of the unequal-variance Gaussian model.

Exercises

9.1. Suppose that the distributions of evidence under signal and
noise conditions are as sketched below. Indicate on the figures where an observer using a likelihood-ratio criterion would place decision criteria. Note:
the specific forms of the distributions are stated for reference, but you
should answer with reference to the figures alone. Calculation is unnecessary.

a. The noise is standard Gaussian $\mathcal{N}(0,1)$ and the signal is $\mathcal{N}(1.4,1)$:

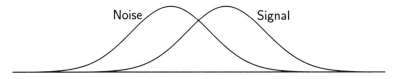

b. The noise is chi-square with three degrees of freedom and the signal is
chi-square with five degrees of freedom:

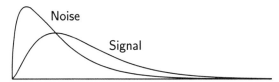

c. The noise is Gaussian $\mathcal{N}(0,2)$ and the signal is a 50–50 mixture of
$\mathcal{N}(-1.5,1)$ and $\mathcal{N}(1.5,1)$ distributions:

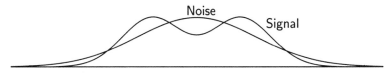

9.2. An observer knows that signal events are twice as likely to occur
as no-signal events. For the distributions in Problem 9.1 indicate where this
observer should place the decision criteria if Bayes' rule is followed.

9.3. Suppose that an observer must decide between two Bernoulli events, one with $\pi = 0.3$ and one with $\pi = 0.24$. Two hundred trials are observed, and the event occurs 54 times.

a. What decision would a likelihood-ratio observer make?

b. What decision would a Bayesian observer make if $\pi = 0.3$ was considered to be three times a likely as $\pi = 0.24$.

9.4. Events occurring randomly in time are sometimes described by a simple probabilistic model known as a *Poisson process*. According to this description, the number of events N observed during a time interval of length t has a *Poisson distribution*:

$$P(N = n) = \frac{e^{-\rho t}(\rho t)^n}{n!}. \tag{9.21}$$

The parameter ρ is the mean rate at which the events occur, measured in events per unit time. Consider two Poisson processes with rates ρ_1 and ρ_2.

a. A process is observed for time t, and n events are counted. Construct a likelihood-ratio (or log likelihood-ratio) rule to decide between the alternatives $H_1 : \rho = \rho_1$ and $H_2 : \rho = \rho_2$.

b. Use this rule to decide between rates of 10 and 15 events per second, based on a count of 30 events obtained during a $2\frac{1}{2}$-second period.

9.5. The mean of the Poisson distribution (Equation 9.21) is $\mu = \rho t$. When the number of events is large, the Poisson distribution has nearly a Gaussian form with variance σ^2 equal to the mean μ. These facts can be used to give a decision procedure for large Poisson variables. Suppose that a Poisson sequence of events is observed for a fixed interval of time. Ordinarily the rate is such that a distribution with a mean of μ_n events is obtained. Sometimes a signal is added that increases the rate of events so that a distribution with a mean of μ_s events is obtained. Thus, the distributions of observations in the two conditions are

$$X_n \sim \mathcal{N}(\mu_n, \mu_n) \qquad \text{and} \qquad X_s \sim \mathcal{N}(\mu_s, \mu_s).$$

The observer must decide whether the signal is present based on a single observation of X.

a. Find the log likelihood ratio for this task, $\log \mathcal{R}(S : N)$.

b. Show by simplifying this ratio that an optimal observer can base a decision on the quantity $y = (\mu_n + \mu_s)x^2 - 2x$.

9.6. A non-Gaussian distribution that arises frequently is the *exponential distribution*—among other places, it appears as the time passed between the members of a series of Poisson events. Suppose that the evidence distributions associated with signal and noise stimuli have the density functions

$$f_n(x) = \tfrac{1}{2}e^{-x/2} \qquad \text{and} \qquad f_s(x) = \frac{1}{2\tau}e^{-x/2\tau},$$

with $\tau > 1$. The corresponding cumulative distribution functions are

$$F_n(x) = 1 - e^{-x/2} \qquad \text{and} \qquad F_s(x) = 1 - e^{-x/2\tau}.$$

a. When signal and noise events are equally likely, where would the optimum likelihood-ratio decision criterion be placed? Plot the distribution and indicate the criterion for $\tau = 4$.

b. Find the algebraic form of the operating characteristic (i.e., Equation 3.2 on page 45). Plot it for $\tau = 8$.

Chapter 10

Multidimensional stimuli

In the basic detection and identification models, the effect of a stimulus presentation is expressed by a univariate random value X. This representation works well for simple stimuli. However, many real stimuli have more that one component, each of which carries some information about the presence or absence of a signal or about its nature. The observer must integrate this information in a single decision. This chapter discusses the signal-detection approach to these multidimensional stimuli.[1]

There are several reasons to study the detection of multicomponent stimuli. First, it is important to understand how a single decision dimension can arise from a complex stimulus. Without this rationale, the appropriateness of the signal-detection principles in many applications is unclear. Second, the relationship between the multicomponent stimuli and the responses can be used to probe an observer's behavior in complex tasks. A comparison of the response of an observer to a multicomponent stimuli with the response to the components alone gives information about the organization of the perceptual system. Third, the multivariate signal-detection model is necessary to understand how an ideal observer can combine multiple sources of information.

10.1 Bivariate signal detection

Most stimuli in the real world are intrinsically multidimensional, with distinct attributes that must somehow be selected and combined into a single decision. This multidimensionality is obvious in something like a face recog-

---

[1]Before proceeding with this chapter, the reader may wish to review the material on joint random variables (page 231) and on the multivariate Gaussian distribution (page 243).

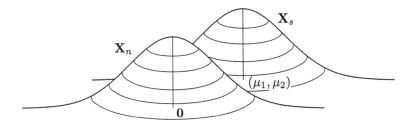

Figure 10.1: Bivariate distributions representing noise and signal events. Compare this picture to that in Figure 1.1 on page 12.

nition study—an observer who must judge whether a picture of a face has been seen before draws on a great many aspects of the picture, yet apparently can integrate these into a single feeling of "familiarity." However, even as simple a stimulus as a pure tone has multidimensional aspects: at the physiological level the tone excites many auditory receptors. It is important, both conceptually and as a part of psychophysical theory, to understand how these disparate components are put together and one decision made.

The signal-detection representation of a multidimensional stimulus is a generalization of the univariate representation. The effect of a single-component stimulus corresponds to a random variable X, and a particular instance of the stimulus is a realization of this random variable. The effect of a two-component stimulus is described by a bivariate random variable $\mathbf{X} = (X_1, X_2)$ and its instances by $\mathbf{x} = (x_1, x_2)$. Instead of being represented as a point on a line, an observation is a point in two-dimensional space. The distribution of $\mathbf{X}$ is expressed by a density function $f(\mathbf{x}) = f(x_1, x_2)$, which associates a density with each point of the space. This density function is characterized by its bivariate mean $\boldsymbol{\mu} = (\mu_1, \mu_2)$ and by various parameters (which depend on the form of the distribution) that express its variability and shape.

The effects of stimuli with more than two components are represented by higher-dimensional random variables—a three-component stimulus by a three-dimensional random variable, and so forth. Instances of these stimuli are points in a space of the appropriate dimensionality. However, although most of the principles that will be discussed apply to d–dimensional stimuli, two-dimensional pictures suffice to illustrate the concepts in this book.

In the basic bivariate detection model, the effects of noise stimuli and signal stimuli are described by realizations of random variables $\mathbf{X}_n$ and $\mathbf{X}_s$. Figure 10.1 shows a pair of bivariate distributions, such as might express the distributions of noise and signal events, one with a mean at $\mathbf{0} = (0, 0)$, the other with a mean at $\mu = (\mu_1, \mu_2)$. These have the Gaussian form that is assumed in most applications of the theory. Depending on the variability relationships, several different versions of the model can be constructed. In the simplest of these models, the two components are independent and the variances σ_1^2 and σ_2^2 are the same for signal and noise stimuli:

$$\mathbf{X}_n \sim \mathcal{N}(\mathbf{0}, \boldsymbol{\Sigma}), \quad \mathbf{X}_s \sim \mathcal{N}(\boldsymbol{\mu}, \boldsymbol{\Sigma}) \qquad \text{with} \qquad \boldsymbol{\Sigma} = \begin{bmatrix} \sigma_1^2 & 0 \\ 0 & \sigma_2^2 \end{bmatrix}.$$

That model is the place to begin.

The problem for the observer who is using multivariate information is to find a rule to reduce the two (or more) dimensions to a single decision. At the center of Figure 10.2, the two bivariate distributions of Figure 10.1 are shown from above by level curves at one standard deviation from the mean. Because the observations on the two dimensions are uncorrelated, these curves are circles with unit radius centered at $\mathbf{0}$ and $\boldsymbol{\mu}$. At the top and left side of the figure are the univariate marginal distributions of X_1 and X_2, ignoring the other variable. With uncorrelated distributions such as these, the marginal distributions have the same variance as the joint bivariate distribution—note that the places where the circle for the noise distribution cuts each axis in the joint representation correspond to points one standard deviation out in the marginal distributions.

As in the univariate situation, the observer's task is to separate observations that come from one distribution from those that come from the other. In univariate signal-detection theory it is easy to see how to map the values of X onto decisions: small values lead to one response and big values to the other. With multivariate stimuli there are many ways to make this mapping. This chapter will describe several rules, differing in their characterization of the stimuli, the task, and the observer.

One approach to the problem is through its geometry. It is clear from Figure 10.2 that the shortest distance between the centers of the two distributions is along the line that connects their means. This line makes a natural decision axis. Each observation $\mathbf{x}$ is projected perpendicularly onto it. Thus, the point $\mathbf{x}$ in Figure 10.2 corresponds to the point on the line between the means that is connected to it by the dashed line. By measuring the signed distance from $\mathbf{0}$ along this line, the bivariate observation is converted to a single number x. The observer bases the decision on this distance. An observation with a projection that is nearer to $\mathbf{0}$ than to $\boldsymbol{\mu}$ is probably a noise stimulus, and one with a projection nearer to $\boldsymbol{\mu}$ than to $\mathbf{0}$ is probably a signal. When the two multivariate distributions are projected onto this axis, they form two univariate distributions. In Figure 10.2 these two distributions are shown on the diagonal axis at the bottom of the figure.

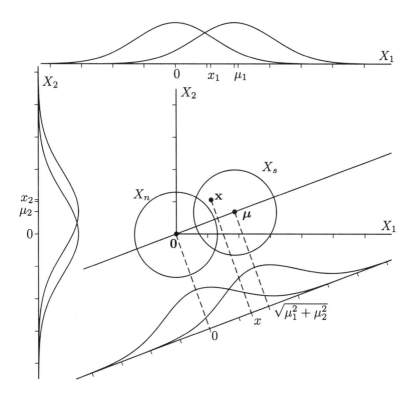

Figure 10.2: *Bivariate signal detection based on the projection of the observations onto the line connecting the means. The components are uncorrelated, with* $\mu = (1.88, 0.69)$. *The three univariate diagrams show single-component detection and detection along the axis connecting the means. The bivariate observation* **x** *projects to the points* x_1, x_2, *and* x *on these axes.*

Decisions using this axis are made exactly as described by the univariate theory of the earlier chapters.

The relationship between the multivariate mean (μ_1, μ_2) and the level of performance implied by the projection rule follows from the geometry. Because the level curves are unit circles, the standard deviation of the projections on any axis is also unity. By the Pythagorean theorem, the distance between means is $\sqrt{\mu_1^2 + \mu_2^2}$. The projections have unit variance, so the joint d' based on the projection is

$$d' = \frac{\text{distance between means}}{\text{standard deviation of projection}} = \sqrt{\mu_1^2 + \mu_2^2}. \qquad (10.1)$$

A bivariate representation comparable to this one was used earlier in this book to analyze position bias in the forced-choice task (Section 6.3). The stimulus on each trial combined a signal instance and a noise instance as a bivariate stimulus, either $\langle SN \rangle$ or $\langle NS \rangle$. A forced-choice response is comparable to discrimination between these two bivariate stimuli. In Chapter 6, it was shown that d'_{FC} for the forced-choice task exceeded d' for simple detection by a factor of $\sqrt{2}$. Exactly the same relationship is expressed by Equation 10.1. The advantage of the forced-choice task over simple detection is a consequence of the bivariate stimulus.

This equal-variance picture is easily modified to describe detection when the components of the signal and noise random variables have different variances. Now the two circles in Figure 10.2 have different diameters. The shortest distance between the distributions is still the line that connects their means, but, unlike the equal-variance case, projections onto that line have variances that depend on the stimulus. The decision process is described either by the unequal-variance Gaussian model of Chapter 3 or by the likelihood-ratio test of Section 9.3. Although the d' statistic is no longer the proper summary of performance, the other measures of Chapter 4 apply.

10.2   Likelihood ratios

Projecting the observations onto the axis that connects the means of the distributions, as in Figure 10.2, no longer makes sense when the components of the multivariate stimulus are correlated or when the multidimensionality includes a mixture of graded and discrete attributes. Either or both of these complications are likely with complex stimuli. The likelihood-ratio tests of Chapter 9 give a more plausible, and in some ways less complex, representation of the multivariate decision process.

The likelihood-ratio procedure and its Bayesian extension apply to multidimensional stimuli exactly as they did to unidimensional stimuli. Each d-component observation is represented by a realization $\mathbf{x}$ of the multivariate random variable $\mathbf{X} = (X_1, X_2, \ldots, X_d)$ with density $f_s(\mathbf{x})$ on signal trials and $f_n(\mathbf{x})$ on noise trials. The observer is aware of the form of these distributions and, when an observation $\mathbf{x}$ is made, estimates the densities (or their logarithms) to obtain the likelihoods

$$\mathcal{L}(N) = f_n(\mathbf{x}) \qquad \text{and} \qquad \mathcal{L}(S) = f_s(\mathbf{x}).$$

These values are compared in a log-likelihood ratio (Equation 9.8):

$$y = \log \mathcal{R}(S:N) = \log \frac{f_s(\mathbf{x})}{f_n(\mathbf{x})} = \log f_s(\mathbf{x}) - \log f_n(\mathbf{x}).$$

As with univariate stimuli, the decision is based on this ratio. A criterion λ is established and the observer reports YES if $\log \mathcal{R}(S:N) \geq \lambda$ and

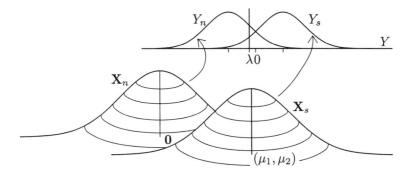

Figure 10.3: *The bivariate sensory distributions $\mathbf{X}_n$ and $\mathbf{X}_s$, and the univariate distributions of the log likelihood ratio, Y_n and Y_s.*

NO if $\log \mathcal{R}(S : N) < \lambda$ (Decision Rule 9.16). The simple likelihood-ratio observer picks $\lambda = 0$, while the Bayesian observer takes the signal frequency into account by calculating the Bayes' ratio or by setting $\lambda = -\operatorname{logit}[P(S)]$ (Equations 9.11 and 9.13).

The effect of the likelihood calculation is to transform any multivariate observation $\mathbf{x}$ into a univariate observation y, either $\log \mathcal{R}(S : N)$ or a monotonic function of it. In this way, the original noise and signal random variables $\mathbf{X}_n$ and $\mathbf{X}_s$ become two new univariate random variables Y_n and Y_s. Figure 10.3 shows schematically the distribution of the four variables. The bulk of the signal distribution lies in the regions where $f_s(\mathbf{x}) > f_n(\mathbf{x})$, and here the variable $y = \log \mathcal{R}(S : N)$ is positive. The bulk of the noise distribution lies in the regions where $f_s(\mathbf{x}) > f_n(\mathbf{x})$, and here y is negative. So the distribution of Y_s lies largely to the right of the origin, while that of Y_n lies largely to the left. A criterion λ applied to the Y axis discriminates the two distributions and determines the decision rule for the multivariate observations. The distributions of Y_n and Y_s in Figure 10.3 and their relationship to the decision criterion are identical to those shown for the univariate decision in Figure 9.1 on page 157. The observer generates a response without further concern about the multidimensional nature of the stimuli.

When the distributions of $\mathbf{X}$ are multivariate Gaussian and the signal and noise variances are the same, the combination that creates Y is a linear combination of the X_j. Moreover, when the component observations are uncorrelated, each dimension is weighted by the magnitude of the evidence that it provides. With equal variance, the density functions (from Equation A.52 on page 244) are

$$f_n(\mathbf{x}) = \frac{1}{(2\pi)^{d/2}} \exp\left(-\tfrac{1}{2}\textstyle\sum x_i^2\right),$$

the sum being over all d components of $\mathbf{x}$, and

$$f_s(\mathbf{x}) = \frac{1}{(2\pi)^{d/2}} \exp\left[-\tfrac{1}{2}\sum(x_i - \mu_i)^2\right].$$

The log likelihood ratio based on these densities is

$$
\begin{aligned}
\log \mathcal{R}(S:N) &= \log f_s(\mathbf{x}) - \log f_n(\mathbf{x}) \\
&= \left[\log \frac{1}{(2\pi)^{d/2}} - \tfrac{1}{2}\sum(x_i - \mu_i)^2\right] - \left[\log \frac{1}{(2\pi)^{d/2}} - \tfrac{1}{2}\sum x_i^2\right] \\
&= \tfrac{1}{2}\sum[x_i^2 - (x_i - \mu_i)^2] \\
&= \tfrac{1}{2}\sum(2\mu_i x_i - \mu_i^2) \\
&= \sum\mu_i x_i - \tfrac{1}{2}\sum\mu_i^2.
\end{aligned}
$$

The first term is the weighted sum of the x_i, and the second term is a constant that centers the origin.

To analyze the properties of this rule, it is easier to drop the second term and take as the decision variable the sum

$$y = \sum\mu_i x_i. \tag{10.2}$$

This linear combination of Gaussian random variables is also Gaussian (page 241). For an observation that came from the noise distribution $\mathbf{X}_n$, the means of the X_{ni} are zero, and $\mathsf{E}(Y_n) = 0$. For signals, the mean is

$$\mathsf{E}(Y_s) = \sum\mu_i \mathsf{E}(X_{si}) = \sum\mu_i^2.$$

The variance of either Y_n or Y_s (Equation A.33 on page 235) is

$$\mathrm{var}(Y) = \sum\mu_1^2 \,\mathrm{var}(X_i) = \sum\mu_i^2.$$

Rescaling the difference in means by the standard deviation gives a d' statistic:

$$d'_{\mathrm{lr}} = \frac{\mathsf{E}(Y_s) - \mathsf{E}(Y_n)}{\sqrt{\mathrm{var}(Y)}} = \frac{\sum\mu_i^2 - 0}{\sum\mu_i^2} = \sqrt{\sum\mu_i^2}. \tag{10.3}$$

This value is the same as that which applied to the distance-based observer (Equation 10.1). The likelihood analysis gives a sounder basis to that procedure and makes clear the special conditions necessary for an optimal observer to use it.

Even when the multivariate random variable $\mathbf{X}$ does not have a Gaussian distribution, the likelihood-ratio model suggests why the Gaussian model may still apply to the final decision. The central limit theorem (page 241) implies that when many independent random variables with similar distributions are added together, their sum approaches a Gaussian distribution. A likelihood-ratio observer detecting stimuli with many independent components more or less satisfies these conditions. The decision is based on the random variable

$$\begin{aligned} Y &= \log f_s(\mathbf{X}) - \log f_n(\mathbf{X}) \\ &= \sum \log f_{si}(X_i) - \sum \log f_{ni}(X_i) \\ &= \sum [\log f_{si}(X_i) - \log f_{ni}(X_i)]. \end{aligned}$$

Each of the bracketed terms is independent of the others. When d is large, the distribution of Y is close to Gaussian unless the sum is dominated by a few terms. This result is more general than the calculation above shows— in particular, independence of the components is not essential. As long as the evidence is truly multidimensional and is not dominated by a few non-Gaussian components, the log likelihood-ratio decision variable has a smooth symmetric distribution that is well approximated by a Gaussian distribution.

10.3   Compound signals

Many natural stimuli are composed of several distinct and separable components. They can be thought of as compound stimuli. Although each of the components can be detected separately, they normally appear in this redundant combination. There are many ways to combine information from several components of a compound stimulus. These rules provide useful benchmarks for judging the performance of a real observer.

Consider a study with three conditions. In two of these, the detectability of two simple signals is measured. In the third condition, the two signals are combined and the detectability of their compound is found. Denote the detectability of the single components by d'_1 and d'_2, and that of the compound by d'_{12}. How the value of d'_{12} relates to d'_1 and d'_2 depends on the way that the observer combines the information provided by the two components.

Even in this simple situation, the number of models for the task is disconcertingly large. Performance depends not only on how the observer combines information from the two components, but also on the relationship between X_1 and X_2 and on any differences in variance (or covariance) between signal and noise. Even an incomplete catalog of the possibilities is large enough to be confusing. This section and the next describe some of the more plausible representations for how the observer can act and their implied relationship between the single-component and the dual-component tasks.

Perhaps the simplest thing that the observer can do with the compound is to concentrate on the most detectable of the signals: if $d'_1 > d'_2$, then attend to the first component; otherwise attend to the second component. Using this *single-look strategy* the detectability of the compound is the maximum of the component detectabilities:

$$d'_{1\text{-look}} = \max(d'_1, d'_2).$$

This strategy may be the only one available when paying attention to one component interferes with the observation of the other. Ignoring the least-detectable component is also a reasonable strategy when the detectability of the two components is very different, so that little information is added by the less-detectable member.

If the observer can combine information from the two components, then the compound signal is generally easier to detect than either of its parts. Performance better than $d'_{1\text{-look}}$ is possible. A second benchmark is established by assuming that both signals are received in noise-free form, and all the performance-limiting variation is introduced by the observer when making the decision. For example, the observer might perceive the signal magnitudes accurately, but be unable to set the criterion consistently from one trial to the next. Without perceptual noise, the observations could be added together without increasing the decision variance. The detectability under this *late-variability model* is the sum of the component detectabilities:

$$d'_{\text{LV}} = d'_1 + d'_2.$$

The process implied by d'_{LV} is not very realistic. Except in very special circumstances, each component has its own intrinsic variability, an assumption that is almost the central premise of signal-detection theory. The two sources of variation make the uncertainty of the combined signal greater than that of either component alone, and the net detectability is less than the sum of the detectabilities of the individual parts.

The two values $d'_{1\text{-look}}$ and d'_{LV} span the range that one would ordinarily expect for d'_{12}. The smaller value describes an observer who can attend only to one stimulus channel at a time; the larger value, an observer who can attend to both perfectly. Values outside this range are possible, however. They suggest that the two parts of the compound signal combine in some rather special way, either destructively when $d'_{12} < d'_{1\text{-look}}$ or constructively when $d'_{12} > d'_{\text{LV}}$. Models that imply performance at these extremes generally involve special properties of the task and the observer, so they are not elaborated here.

There are many ways in which the observer can combine information from the two components that give performance between $d'_{1\text{-look}}$ and d'_{LV}. A slightly more sophisticated version of the single-look strategy is to test each stimulus component separately and combine the results. A signal is deemed present by the *dual-look strategy* if either $X_1 > \lambda_1$ or $X_2 > \lambda_2$, for appropriate criteria λ_1 and λ_2. This approach does not lead to a simple expression that relates the combined performance to the individual d'_1 and d'_2 (the criteria are also involved). However, when the components are independent, the probabilities of misses and correct rejections are the product of their component counterparts—for example, a compound signal is missed whenever neither X_{1s} nor X_{2s} exceed their respective criterion. Using single subscripts to indicate the probabilities for components presented alone and a double subscript for the dual-look outcome, the probability of this joint event is

$$(1 - f_{12}) = (1-f_1)(1-f_2).$$

Thus, the false-alarm and hit probabilities for the dual-look strategy are

$$\begin{aligned}
f_{12} &= 1 - (1-f_1)(1-f_2) = 1 - \Phi(\lambda_1)\Phi(\lambda_2), \\
h_{12} &= 1 - (1-h_1)(1-h_2) = 1 - \Phi(\lambda_1 - d'_1)\Phi(\lambda_2 - d'_2).
\end{aligned}$$

(10.4)

From these values the combined measure of detectability $d'_{2\text{-look}}$ can be calculated (see Problem 10.2).

An ideal observer would not use the dual-look strategy. Superior performance is possible using a likelihood ratio. The analysis in the last section gives the level to expect when each component of the compound stimulus brings independent information to bear on the detection. From Equation 10.3 with $\mu_1 = d'_1$ and $\mu_2 = d'_2$,

$$d'_{lr} = \sqrt{(d'_1)^2 + (d'_2)^2}.$$

(10.5)

This is the largest value that is possible when the components are independent and their noise sources are distinct.

Example 10.1: An observer is presented with blocks of trials of three types. Two are simple detection, one of a soft tone, the other of a dim light. Performance statistics for these single-modality conditions are $d'_{tone} = 1.1$ and $d'_{light} = 1.3$. The third task is detection of a compound of tone and light. What performance could be expected for this task?

Solution: The poorest performance that would be expected here would occur if the observer attended to one signal alone, but chose the wrong one. Then $d'_{\text{1-look}} = d'_{\text{tone}} = 1.1$. A slightly better observer would pick the light, giving $d'_{\text{1-look}} = d'_{\text{light}} = 1.3$. As the signals are in different modalities, they plausibly act independently. The performance of a dual-look observer depends on where the criteria are set. Suppose that they were set at $\lambda_1 = 0.9$ and $\lambda_2 = 0.8$. Then from Equations 10.4,

$$f = 1 - \Phi(0.9)\Phi(0.8) = 0.3570,$$

$$h = 1 - \Phi(0.9 - 1.1)\Phi(0.8 - 1.3) = 0.8703.$$

Using the standard formula for estimating d' from these two proportions (Equation 2.4 on page 24), $d'_{\text{2-look}} = 1.49$. A likelihood-ratio observer does somewhat better (Equation 10.5):

$$d'_{\text{lr}} = \sqrt{(1.1)^2 + (1.3)^2} = \sqrt{2.90} = 1.70.$$

To the extent that the actual observer is not ideal, performance falls below this value; to the extent that the original performance is limited by nonsensory factors, performance might be higher. Under the unrealistic supposition that performance was limited only by the decision process, $d'_{\text{LV}} = 2.4$ could be reached.

One way to illustrate many of the different multivariate decision rules (although not every one) is by drawing the space containing the distributions and indicating which combinations of x_1 and x_2 are assigned to which outcomes. This operation defines a *decision bound* that separates the space into regions associated with each response. The observer acts as if this bound were used to map observations onto responses.

Figure 10.4 shows the decision bounds for four rules. Bivariate distributions associated with the stimuli are shown as circles with one standard deviation radius centered at the means $(0, 0)$ and $\mu = (d'_1, d'_2) = (1.2, 1.7)$. The decision bounds are shown as dashed lines. The bound for the single-look rule is a horizontal line (labeled "1-look") that separates the lower part of the figure from the upper. The horizontal orientation of this bound indicates that no attention is paid to the X_1 component of the compound. The bound for the dual-look rule (labeled "2-look") is formed from two lines at right angles, cutting off the lower left corner of the space. Because the two limbs of this bound are horizontal and vertical, each part of the decision is based on a single component of the stimulus.

Rules that involve integration of information from the components before the decision have bounds that are not parallel to the axes. Rules based on linear combinations of the components are represented by straight lines. Two of these are shown: one for a simple sum, the other for the likelihood ratio. The bound labeled "Sum" describes a criterion based on

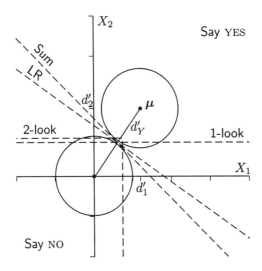

Figure 10.4: Decision bounds for four strategies for detection of the compound signal with $\boldsymbol{\mu} = (1.2, 1.7)$. See text for explanation.

$y = x_1 + x_2$. Because the two components are not equally informative, this rule is less than optimal when $d'_1 \neq d'_2$. The optimal likelihood-ratio rule (labeled "LR") is equivalent to one based on the weighted combination $y = \mu_1 x_1 + \mu_2 x_2$ (Equation 10.2). This tilted line most completely separates the two distributions.

Some decision procedures cannot be represented by decision bounds. Bounds cannot describe any rule that does not assign a unique response to each point in the (X_1, X_2) space. In particular, the description that led to the late-variability d'_{LV} does not correspond to a decision bound. That level of performance implies that the variability of the responses is not associated with the perceptual processes, but with the decision process. A given observation, say one falling roughly midway between $\mathbf{0}$ and $\boldsymbol{\mu}$, will sometimes receive one response and sometimes another. Another decision process that cannot be represented by a decision bound is a weaker variant of the single-look strategy. The observer does not know which component is the larger, so on each trial haphazardly chooses one of them with probability $1/2$, and responds only to it.

It is frequently found that several different descriptions of the observer imply the same relationship between d'_1, d'_2, and d'_{12}. For example, with the equal-variance distributions in Figure 10.4, a decision based on the likelihood-ratio bound is equivalent to one based on the projections onto the line between means, as in Figure 10.2. For these observers, $d'_{\text{proj}} =$

$d'_{1r} = \sqrt{(d'_1)^2 + (d'_2)^2}$. Moreover, if the components have equal individual detectability, then these procedures are equivalent to that of an observer who sums the components and bases the response on $y = x_1 + x_2$. Although their implications are the same, the processes by which the decision is made are different in the three cases. The projection observer is manipulating the space of observations, the likelihood-ratio observer is judging the relative likelihoods of the observation, and the summation observer is adding the two stimulus components. The discovery that $d'_{12} = \sqrt{2}d'$ is not sufficient to sort out these possibilities. Because of equivalences such as these, one must be very careful when trying to infer process from empirical data.

10.4   Signals with correlated components

The analysis in Section 10.3 depends on the independence of the two component signals. The level curves that express the variability of such stimuli are circles. The variance is the same along any axis through these distributions, and the distance between **0** and μ can be used as a measure of detectability. However, in many situations, the information carried by the two signals is not independent. For example, the components of the multivariate observations might be the output of two (or more) physiological receptors with partially overlapping sensitivities. A physical stimulus stimulates, to some degree, both receptors, so that X_2 is larger when X_1 is large than when it is small and vice versa. This association between the signals is expressed by the correlation ρ between the components. The greater the correlation, the less the information provided the pair of observations exceeds that provided by one alone and the smaller is the advantage of using both components. When the correlation is perfect (i.e., when $\rho = 1$), the components are completely redundant, and the multivariate observation is no more effective than the univariate.

Correlation of the responses brings a new factor to the relationship between the detection of simple signals and that of their compound. It is no longer reasonable to assume that the mean evidence provided by a given component is the same for single-component and dual-component stimuli. Whatever aspect of the situation induces the correlation between X_1 and X_2 can also shift the center of the distribution. As a result, the statistic d' obtained by measuring detectability of signal S_1 when signal S_2 is absent can differ from the displacement of the mean of X_1 in the compound signal. The single-signal experiments compare random variables centered at $(0, 0)$ and $(d'_1, 0)$ and at $(0, 0)$ and $(0, d'_2)$ in the multivariate space, while the compound-signal experiment compares random variables centered at $(0, 0)$ and (μ_1, μ_2). If the presence of signal S_2 also increases the size of X_1, as

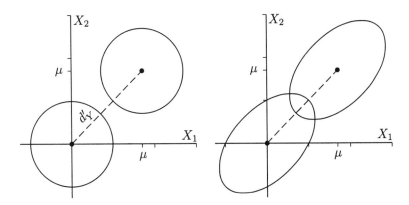

Figure 10.5: The combination of signals with univariate means $\mu = 1.7$ when the components are independent (left) and correlated with $\rho = 0.5$ (right).

it would with the overlapping receptions mentioned above, then $\mu_1 > d'_1$. These possibilities cannot be sorted out by looking at d'_1, d'_2, and d'_{12} alone. The assumption that the means in the two studies are unrelated is more reasonable with uncorrelated signals than it is with correlated signals.

If one puts this caveat about means to the side, the effects of the correlated signals on the detection is easy to represent in a distributional diagram. Figure 10.5 illustrates the situation when both signals have mean strength $\mu = 1.70$ in the compound. In the left panel, the two stimuli are uncorrelated. The variance along the diagonal is the same as that along either axis, and the diagonal distance of $\sqrt{2}\mu = 2.40$ is directly interpretable as d'_Y. In the right panel, the components are correlated and the contours are ellipses. The variability associated with each component alone is the same as in the uncorrelated figure—note that the contour for the noise stimulus crosses the axes at the same place in both panels. However, the ellipses are elongated in the direction that separates the two stimuli, indicating that the variablity in that direction is larger than along either axis. The distance between the centers must be divided by this greater variability to get d'_{12}.

A likelihood-ratio analysis determines the effect of correlation on detectability for the ideal observer. When the observations have correlated Gaussian distributions with the same mean and variance for both components, the optimal decision variable is the sum of the components, $y = x_1 + x_2$. A proof of this fact (not given here) follows the derivation of Equation 10.2, but is done with matrices. To determine the detectability, note that the means are not changed by the correlation, but that the variance is. The variance of the sum of two random variables X_1 and X_2 that are correlated with correlation ρ is

$$\text{var}(X_1 + X_2) = \text{var}(X_1) + \text{var}(X_2) + 2\rho\sqrt{\text{var}(X_1)\,\text{var}(X_2)}$$

(Equation A.34 on page 235). When both original variables have unit variance, $\text{var}(Y) = 2(1+\rho)$, which is larger than the variance of the sum of uncorrelated variables by a factor of $1 + \rho$. As a result,

$$d'_{\text{lr}} = \frac{E(Y_s) - E(Y_n)}{\sqrt{\text{var}(Y)}} = \frac{2\mu - 0}{\sqrt{2(1+\rho)}} = \frac{\sqrt{2}\,d'}{\sqrt{1+\rho}}. \tag{10.6}$$

When the signals are independent, $\rho = 0$, and Equation 10.6 becomes $d'_{\text{lr}} = \sqrt{2}d'$, as in Equation 10.5. When the signals are completely redundant, $\rho = 1$, giving $d'_{\text{lr}} = d'$, and there is no improvement in performance. Intermediate correlations fall between these limits. They give the optimal performance possible with the available information.

> **Example 10.2:** An unbiased observer correctly detects each of two signals 75% of the time in a study with equally frequent signals and noise stimuli. The compound stimulus formed from the two signals is detected 81% of the time. Estimate the correlation between the signals. To what extent do the components overlap?
>
> *Solution:* To solve this problem, one must assume that forming the compound does not change the mean effects, so that d' obtained from the first studies can be taken as μ in the third. If misses and false alarms are equally frequent, then the 75% correct-response rate indicates that the criterion lies 0.674 units above the noise mean and the same distance below the signal mean. Hence, $d'_1 = d'_2 = 2 \times 0.674 = 1.35$. If these values are taken as the means of the compound, then performance on it could be about $\sqrt{2}d' = 1.91$. However, the 81% level implies that $d'_{12} = 2 \times 0.842 = 1.68$. Putting $d' = 1.35$ and $d'_{\text{lr}} = 1.68$ in Equation 10.6 and solving for the correlation gives
>
> $$\sqrt{1+\rho} = \frac{\sqrt{2}\mu}{d'_{\text{comp}}} = \frac{1.91}{1.68} = 1.137.$$

Thus, $\hat{\rho} = 0.29$. The two signals have about $\hat{\rho}^2 = 9\%$ of their variability in common.

Equation 10.6 implies that when the correlation between signals is negative, the compound signal should be more detectable than it would be were the components independent. Although that interpretation is correct, the

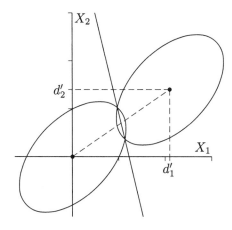

Figure 10.6: *The likelihood-ratio decision bound (solid line) for correlated signals with unequal detectability.*

situation does not arise often. It is easier to physically realize situations in which shared aspects of the components give them a positive correlation than a negative correlation. Performance typically falls at or below $\sqrt{2}\mu$, not above it.

If the signals have different strengths or if the variability of signal and noise differs across components, then summing the components is not optimal. When the strengths of the signals are unequal and the correlations are the same for both stimuli, the optimal decision can still be still based on a linear combination of the observations, but the coefficients are not the univariate detectabilities (as in Equation 10.2). The weights for the optimal decision variable depend on all the parameters of the distribution:

$$y = (d_1' - \rho d_2')x_1 + (d_2' - \rho d_1')x_2, \qquad (10.7)$$

or, in matrix terms, $y = \boldsymbol{\mu}'\boldsymbol{\Sigma}^{-1}\mathbf{x}$. Figure 10.6 shows the decision bounds when $\rho = 0.5$, $d_1' = 2.1$, and $d_2' = 1.4$. The axes of the ellipses do not line up with the line connecting the means, as they did in Figure 10.5. The optimal decision bound is not perpendicular to this line. Projection, in the sense illustrated in Figure 10.2, is onto an oblique axis. Calculations of this type are important in branches of multivariate statistics, where distances measured with respect to this axis are known as the *Mahalanobis distance*.

When the variances are different for signal and noise, or when the correlations between the signals depend on which stimulus was presented, the picture is further complicated. The shapes or inclinations of the ellipses representing the variability need not be the same. The likelihood-ratio decision principle still applies, however. An observer can maximize performance, even in this situation, by selecting the stimulus that has the largest density. The likelihood-ratio analysis of the unequal-variance model for a unidimensional stimulus (Section 9.3) showed that the optimal decision variable was a quadratic in x. The same applies here. In the two-dimensional space, the decision bounds are quadratic curves, such as hyperbolas, parabolas, ellipses, or circles.

The configurations described in the last two paragraphs are considerably more complex than those developed earlier. They would rarely be fitted to simple yes/no detection data (although they might be used with an appropriate type of rating-scale data). They would normally be derived from a particular theory or stimulus configuration. These situations are well outside the scope of this book. However, it is important to realize that the likelihood-ratio principle provides a general approach, even to these problems.

10.5   Uncertainty effects

The last few sections have described how multivariate information is combined in a compound stimulus. Another application of the multidimensional model is to situations in which there is uncertainty about the signal. In a detection task with uncertainty, noise stimuli are presented on a certain proportion of the trials and signals on the others. As in ordinary detection, the observer must distinguish between a signal and its absence. However, unlike the simple detection experiment, the signals are not all identical. There is a set of potential signals, and, on each signal trial, a randomly chosen member of this set appears. The observer does not know which of the signals (if any) will be presented and does not need to identify it if it is (although that may be part of an ancillary task). Only a YES/NO decision is required. The surprising thing about this task is that it is usually harder than when the observer is told which signal to expect. The signal-detection model lets one both understand what makes the uncertain task harder and calculate the extent to which performance declines.

Situations in which the signal is not known exactly are common, particular in practical applications. For example, a radiologist who is examining a film for an abnormality is beset by at least two sorts of uncertainty. Any one of several types of abnormality may be present (a signal-type uncertainty), and it may appear at any of several places on the film (a location uncertainty). Uncertainty can be introduced experimentally in more controlled environments. An observer listening for a tone may hear either noise or noise plus a tone, but that tone may be at any several frequencies (a signal-type uncertainty) or it may originate from any of several spatially distinct places in the room (a location uncertainty). Similarly, an observer watching a screen for a very faint patterned signal may be presented with one of several patterns in one of several locations. In any of these situations, the stimulus is either the noise event N or one of t signals, S_1, S_2, ..., S_t. The task is to respond NO to N and YES to any of the S_i, without regard to which it is.

One would like to relate detection when the signal is uncertain to that when the signal is known in advance. A naive analysis suggests that the uncertainty has no effect on the detection. In terms of the physical events, a trial when signal S_i is presented in the uncertain-signal study is just like that in a simple detection study using S_i, and so it would seem that its detectability should not change. Values of d' calculated by comparing these trials to noise trials should be just like those obtained from known-signal trials. However, this explanation is not correct (except, strictly, for a late-variability representation in which the only source of variability arises from the observer's responses). Even though the received signal does not change, the tasks differ because the observer cannot use the same decision rule. Faced with signal uncertainty, even an ideal observer performs more poorly than when the signal is known. This decrement is not produced by some sort of interaction among the signals or by an attentional effect, but, as the following analysis shows, is directly a result of the uncertainty.

The representation of uncertainty in the signal-detection framework is shown in Figure 10.7 when there are two possible signals with equal and independent effects. A noise stimulus generates an observation of a bivariate Gaussian distribution centered at the origin $\mathbf{0}$. The signal distribution is a mixture of two distributions, one for signal S_1, the other for signal S_2. Because the signals are independent, their associated distributions are located along two distinct axes, one at $\boldsymbol{\mu}_1 = (d_1', 0)$ and the other at $\boldsymbol{\mu}_2 = (0, d_2')$. The observer must create a decision bound that separates the noise distributions from a mixture of these two signal distributions.

The dashed lines in Figure 10.7 show four decision bounds, applicable to different observers or different tasks. An observer who knows that signal S_1 is to be presented ignores the second dimension and uses a decision bound that separates the distribution at $\mathbf{0}$ from that at $\boldsymbol{\mu}_1$. The vertical bound between these distributions is optimal. Similarly, an observer who knows that signal S_2 is to be presented uses the horizontal decision bound between $\mathbf{0}$ and $\boldsymbol{\mu}_2$. Each of these bounds treats the other signal no differently from the noise. The observer in the uncertain-signal case cannot use these decision bounds, but must adopt a rule that separates the noise distribution from the mixture of the two signals. Two such signal bounds are shown in the figure. One is the diagonal line implied by a simple combination rule; the other, a curved boundary implied by a likelihood-ratio criterion.

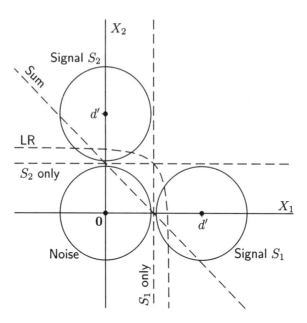

Figure 10.7: *The effects of uncertainty on detection. The noise distribution is centered at the origin; the signal distribution is a mixture of two distributions displaced (with $d' = 2.1$) along orthogonal axes. Four decision bounds, described in the text, are shown by dashed lines.*

Consider the simpler of these bounds first. On any trial, a bivariate observation $\mathbf{x} = (x_1, x_2)$ is made. The observer knows that one of the two components of this vector may be incremented, but not which one. To get a single measure of evidence, the two observations are summed, and the decision is made by applying a criterion to the result. As it did in Figure 10.4, this rule gives a decision bound that is a diagonal line. Because this line must cut off both signal distributions, it passes closer to the center of all three distributions than do either of the known-stimulus lines, implying poorer performance.

The properties of this decision rule are easily analyzed. Geometrically, the 45° diagonal criterion line is closer to the centers of the distributions than the single-criterion lines by a factor of $\sqrt{2}$. Thus, d'_{unc} for the uncertain case equals $d'/\sqrt{2}$. Performance is reduced to about 71% of its former value. More formally, the decision is based on the random variable $Y = X_1 + X_2$. For the noise stimulus, the mean is zero and the variance is the sum of the variance of the component variances, so that $Y_n \sim \mathcal{N}(0, 2)$. For signal stimuli, either X_1 or X_2 has a mean of d', and the other component has a mean of zero, giving $Y_s \sim \mathcal{N}(d', 2)$. A d' statistic calculated from these distributions gives the same result as the geometric argument:

$$d'_{\text{unc}} = \frac{\mathsf{E}(Y_s) - \mathsf{E}(Y_n)}{\sigma_Y} = d'/\sqrt{2}.$$

This model for decisions under uncertainty generalizes to tasks with $t > 2$ alternatives. The decision is based on the sum of all components. The means of Y_n and Y_s remain the same as they were for two alternatives, but the variability increases with the number of terms in the sum to $\sigma_Y^2 = t$. As a result, the detectability declines as the square root of the number of alternatives:

$$d'_{\text{unc}} = d'/\sqrt{t}. \tag{10.8}$$

This relationship gives a rough picture of the increased difficulty created by an uncertain signal.

The size of the reduction in detectability implied by Equation 10.8 is correct only for independent signals of the same strength. When the signal alternatives are not equally detectable, the sum of the components does not give the optimal linear combination. The best straight-line decision bound is tipped to give greater weight to the better-detected signal. Calculations are more complicated as well. The performance must be determined by accumulating hits and false alarms separately before calculating d'_{unc} (see Problem 10.3).

Another restriction to Equation 10.8 is the assumption of independence. It implies that the alternative signals differ in fundamental ways. When the signal alternatives have much in common, as they would when the uncertainty resulted from small variations in a signal stimulus, the reduction in detectability is much smaller.

The use of a linear combination to make the decision is not optimal. A likelihood-ratio decision rule is superior. The likelihood-ratio bounds are conceptually simple. The noise distribution is a single Gaussian, but the signal distribution is bimodal. In the upper part of the space, the signal S_2 dominates S_1, and the response bound looks something like the horizontal line that separates S_2 from N. On the right of the space, S_1 dominates, and the bound is near vertical. More formally, the density function in the signal condition is a mixture of t density functions:

$$f_s(\mathbf{x}) = \frac{f_1(\mathbf{x}) + f_2(\mathbf{x}) + \cdots + f_t(\mathbf{x})}{t}.$$

The ratio of this average to the noise density gives the likelihood function. As always, performance is optimal, in the sense of fewest errors, when this ratio is one, that is, when $f_n(\mathbf{x}) = f_s(\mathbf{x})$. This equation is solved to obtain the optimal decision bounds. The bound, labeled "LR" in Figure 10.7, bends to separate the three distributions more completely than is possible with any straight line. More details are in Problem 10.4.

Calculating the level of performance implied by the likelihood-ratio model is made more difficult by the mixture distribution. The density function must be integrated over an irregular region, a calculation that usually must be done numerically. This analysis shows that performance is somewhat improved over that based on the summation strategy, although it falls short of the performance when the form of the signal is certain. However, much of the difference between the Sum and LR bounds falls in regions where both distributions have little density. The reduction by a factor of $\sqrt{t}$ based on a linear combination (Equation 10.8) is an adequate approximation and a good heuristic for understanding uncertainty effects.

The sum and likelihood-ratio decision rules are not the only strategies that an observer might use. The shape of the likelihood-ratio decision bound suggests an approach based on two univariate decisions, like the dual-look strategy for a compound signal. First, the observer compares X_1 and X_2 and selects the larger of the two, on the grounds that it is most likely to derive from a signal, if one were present. The smaller value is ignored. Second, a criterion is applied to this maximum to make the decision between noise and signal. The bounds implied by this rule are formed by two straight lines, one horizontal and one vertical, placed so as to cut off the noise distribution in the lower-left quadrant. These lines are somewhat farther from the noise distribution than are the single-signal criteria in Figure 10.7. Together, they closely approximate the likelihood-ratio bound. As suggested by this similarity, detection using this rule is only slightly worse than optimal.

Reference notes

In detection theory, the multivariate model was originally developed for identification (see the Tanner reference cited in Chapter 7), although its statistical antecedents are much older. Macmillan and Creelman (1991) develop the models here further and describe some additional models. The different forms of independence and dependence with compound stimuli are discussed by Ashby and Townsend (1986). This work has led to the development of various decision-bounds models; for some examples see the papers collected in Ashby (1992). Distinguishing the position of the means for a compound signal from those for a simple signal requires a study that measures the detectability of both components; see Wickens and Olzak

(1992) in the Ashby volume. The Mahalanobis distance is described in any multivariate statistics text.

Exercises

10.1. A researcher adjusts the intensities of two signals until each of them can be detected about 80% of the time.

a. If these signals are independent, what is the largest value of d' to expect when they are combined as a compound stimulus? Assume all variability is associated with the signals.

b. Under similar assumptions, what is the maximum value d' when the two signals are used in an uncertainty task?

c. In what direction should each of these values change if the signals are not independent, but share qualities?

10.2. Suppose that on observer who is using the dual-look strategy to detect a bivariate compound stimulus puts the decision criteria at $\lambda_i = \frac{1}{2}d'_i$, where the d'_i are the signal-stimulus detectabilities.

a. Calculate the hit rate, the false-alarm rate, and d' for this observer using the signals in Example 10.1. What is the probability that this observer makes an error?

b. Why is this observer less than optimal? In what direction should the criteria be moved? Hint: consider the hit and false-alarm rates, and compare the error probability to that for the criteria in Example 10.1. What would happen to these values as the number of components increases?

10.3. Suppose that signal S_1 is detected with d'_1 in a single-signal study and that signal S_2 is detected with $d'_2 \neq d'_1$ in a comparable study. When the signal is uncertain, the observer bases the decision on the sum of the components.

a. Determine the hit and false-alarm rates for each of the signals and average them to get the hit and false-alarm rates for an uncertain signal. Use these values to calculate the d' that would be observed.

b. Suppose that $d'_1 = 1.2$ and $d'_2 = 1.9$. What performance is predicted for this observer in the uncertain-signal experiment?

10.4. Findixlikelihood-ratio observer!uncertainty effects the decision bounds for a likelihood-ratio observer in the uncertain-signal task shown in Figure 10.7. Assume that the signals are independent with individual detectabilities of d'_1 and d'_2 and the noise and signal variances are equal.

a. Write the densities $f_n(x_1, x_2)$ and $f_s(x_1, x_2)$ using the univariate Gaussian density $\varphi(z)$.

b. Find the likelihood ratio, and show that

$$\mathcal{R}(S:N) = \frac{1}{2}\left[\frac{\varphi(x_1 - d_1')}{\varphi(x_1)} + \frac{\varphi(x_2 - d_2')}{\varphi(x_2)}\right].$$

c. Substitute the densities to get

$$\mathcal{R}(S:N) = \frac{1}{2}\{\exp[d_1'(x_1 - \tfrac{1}{2}d_1')] + \exp[d_2'(x_2 - \tfrac{1}{2}d_2')]\}.$$

d. The unbiased likelihood-ratio observer chooses $\mathcal{R}(S:N) = 1$. Make this assignment and solve for x_2 to get the equation of the decision bound:

$$x_2 = \tfrac{1}{2}d_2' + \frac{1}{d_2'}\log\{2 - \exp[d_1'(x_1 - \tfrac{1}{2}d_1')]\}.$$

10.5. Suppose that noise events are represented by a standard uncorrelated bivariate Gaussian distribution centered at $(0, 0)$ with unit variance, and that signal events are represented by a uncorrelated distribution, also centered at the origin, with variance $\sigma^2 > 1$.

a. What form does the decision boundary take?

b. What is the simplest function of the observation (x_1, x_2) that a likelihood-ratio observer could use? What distribution will it have? Note: the distribution is the one whose properties were examined in Problem 9.6.

Chapter 11

Statistical treatment

The parameters of the signal-detection theory model are theoretical entities and cannot be observed directly. To apply the theory, one estimates these parameters from data and tests hypotheses about their values. Sometimes, as with the estimates of d' and λ for the equal-variance Gaussian model, the estimation procedure is natural and easy, being only a transformation of the original data. At other times, as with the rating-scale model, the estimation requires more elaborate statistical methods. This chapter describes the statistical models and tests that are appropriate for signal-detection models.

11.1   Variability in signal-detection studies

The results of two experiments or conditions within an experiment are rarely exactly the same. Even identical replications of an experiment yield superficially different results. There are many sources for these differences, some systematic and some random. It is important in scientific research to separate the systematic effects from any unsystematic sources of variability. One wants one's conclusions to transfer reliably to other individuals and other circumstances, but does not want to make too much of differences that are purely accidental. There are many ways to eliminate the accidental effects and focus on the systematic ones, some procedural and some statistical. A helpful way to begin a discussion of the statistical aspects of detection theory is to examine the various influences that make one observation different from another.

Consider a simple tone detection experiment. On one trial an observer responds YES, on another trial NO. Why are these responses different? Among the many reasons, several classes are important to the statistical treatment. One possibility is that the stimulus has changed, that the signal was present on one trial and not on the other. However, a change of stimulus is not the whole story—otherwise, there would be no need for signal-detection theory. There are other influences on the responses.

- *Interobserver differences.* One observer differs from another. One observer may be more sensitive to the signal than another, with a higher proportion of hits and a lower proportion of false alarms. Similarly, biases differ among individuals; one may favor YES responses and another NO responses.

- *Intraobserver performance changes.* Observers change over trials. As an observer learns a detection task and becomes skilled at it, performance generally increases. Working against this trend, performance declines toward the end of a long session as the observer becomes fatigued. The probability that an observer reports YES to a signal increases with practice and declines with fatigue. The NO responses to noise follow a similar pattern. Observers may also exhibit other shifts as they change their strategy for making the detection. Another source of intraobserver variability is introduced when a study extends over several days. The observer may be more alert on one day than on another, making the sessions differ.

- *Sequential changes.* The observer's response on one trial may be influenced by the response that was made on the other trials. For example, many observers resist making long strings of responses of one type— the more YES responses in a string, the more likely the observer is to say NO the next time.

- *Residual variation.* There are trial-to-trial fluctuations that cannot be accounted for by any of the above factors. Some of these are undoubtedly generated by the subject, who may vary in attentiveness, receptivity to the signal, decision bias, and so forth, in a completely unpredictable manner. Other fluctuations derive from the signal itself. This type of variability is clearly intrinsic to complex stimuli such as photographs, but is present for simpler stimuli as well. For example, the physical properties of a tone will vary slightly from one presentation to another, as does the ambient situation in which the observation takes place. Careful control—good sound equipment, a quite room, a practiced observer, and the like—can reduce this variability, but cannot altogether eliminate it.

Each of these sources of variation reduces the reliability of any single observation.

The four sources of variability are not all treated in the same way. The last source—the nonremovable residual variation—is the most fundamental to signal-detection theory. The random variables X_n and X_s of the detection model express this type of variability. The greater portion of this book treats this class of variation. That emphasis continues into this chapter. The uncertainty introduced by the residual variation can itself be modeled, and its characteristics are described in Section 11.2. From that analysis, the uncertainty in the estimates of the model's parameters can be worked out. Section 11.3 describes the implications of this variability on the estimated parameters of the equal-variance Gaussian model, and Sections 11.4–11.6 describe the statistical treatment of data based on these estimates. They are the best-case values for the uncertainty.

The other three sources of variability fall outside the basic signal-detection models. Neither interobserver nor intraobserver variability in the detection parameters, nor serial effects across the trials are part of the simple model. As much as possible, most researchers try to minimize the influence of these sources. The traditional way to minimize parameter variability has been to use a few highly trained observers, each examined carefully. Much psychophysical research uses this format.

An experiment of any complexity conducted on a single observer requires many observations to be made. For example, at least three pairs of hit and false-alarm rates are needed to evaluate the unequal-variance Gaussian model and estimate its parameters by the methods of Chapter 4. Determining each of these proportions with sufficient accuracy requires many hundreds of trials. Although obtaining confidence ratings reduces the number of trials required, the complete experiment often must be run over several sessions and several days. In such an experiment, intraobserver sources of parameter variability, particularly practice effects, are of concern. These effects can only be overcome by using observers who have already reached stable performance levels before the start of the experiment. The number of practice sessions needed for the observer's performance to asymptote further extends the study. Even after much practice, there is likely to be some variation among sessions. This variability raises the problems of aggregation mentioned in Section 4.6. These issues are treated statistically in Section 11.7.

In studies that use only a few extensively studied observers, the importance of interobserver comparisons is minimized. The generality of a finding over observers is demonstrated (if at all) by independently analyzing two or three observers, separately presenting each observer's properties and showing that each leads to the same conclusions, then arguing that other observers would behave more or less similarly. This approach is quite reasonable, particularly when interobserver differences are deemed less likely or less important and when much training is necessary to obtain stable behavior. In some domains, however, such studies are either impractical or inappropriate. Sometimes it is impossible to obtain observers for long enough to attain the necessary precision, either because observers are only available for a limited period of time (as in studies that use the typical undergraduate student population), or because only a limited pool of material is available (as in some studies with linguistic materials). The experimenter must substitute smaller amounts of data from a greater number of observers. Interobserver variability is an important component of these studies.

The presence of interobserver variability is not necessary a liability, however. The goal of many investigations is to generalize over a population of observers, and it is difficult to obtain two or three observers who would be universally regarded as representative of this population. The multiobserver design is necessary here in order to obtain good estimates of the extent to which the population of observers is heterogeneous and to which differences between conditions or groups exceed this heterogeneity. Much psychological research is at this level. The statistical approach to these studies is quite different from the approach to designs in which the residual variability dominates. These methods are discussed in Section 11.7.

11.2   Fundamental sampling distributions

Just as the observation of a single detection trial is an imperfect indicator of the true nature of the stimulus, the outcome of an experiment is an imperfect indicator of the true nature of what an observer is doing. The results of two otherwise-identical experiments usually differ. From this point of view, it makes sense to treat as a random variable any quantity calculated from the data, such as the proportions f, h, the correct-response proportion c, and the parameter estimates $\widehat{d'}$, $\widehat{A_z}$, $\widehat{\lambda}$, and $\widehat{\log \beta}$. The distributions of these random variables are induced by the variation in outcomes over experiments and are known as *sampling distributions*. They are central to all statistical analysis.

The sampling distribution of a particular quantity is created by the probabilistic process that generates it. If one knows the probabilistic structure of the observations, then one can derive the sampling distribution theoretically. These theoretical derivations allow one to determine how likely a particular outcome is to occur and to discriminate systematic and reproducible effects from those that are likely to be accidental.

The fundamental sampling distribution in signal-detection theory is that of the responses made to each type of stimulus. In a detection task, this distribution is of the combination of false alarms and hits (or, redundantly, of correct rejections and misses). Specifically, for an experiment in which N_n noise trials and N_s signal trials are presented, the uncertainty of all later measures depends on the joint distribution of the number of false alarms and hits, x_f and x_h.

The sampling distribution of the pair (x_f, x_h) cannot be obtained without making some assumptions about the observations. The most useful analysis is based on two assumptions:

1. The observations are *identically distributed*; that is, the probability of a particular response to a particular stimulus is constant.
2. The observations are *independent*; that is, the probability of a particular response on one trial is unaffected by the responses made on other trials.

With these assumptions, the importance of which will be discussed at the end of this section, the desired probability distribution can be found.

The distribution of a set of discrete independent and identical events is well known. Consider the noise trials. On each trial the observer makes one of two responses, YES or NO, so the trial is either a false alarm or a correct rejection. The probability of a false alarm, derived from the theoretical signal-detection model is P_F (e.g., Equation 1.1). In the N_n trials, the number of events with this probability is given by the binomial distribution (Equation A.35 on page 236):

$$P(x_f; N_n, P_F) = \frac{N_n!}{x_f!(N_n - x_f)!} P_F{}^{x_f}(1 - P_F)^{N_n - x_f}.$$

Similarly, the distribution of the number of hits on signal trials is

$$P(x_h; N_s, P_H) = \frac{N_s!}{x_h!(N_s - x_h)!} P_H{}^{x_h}(1 - P_H)^{N_s - x_h}.$$

The full experiment includes both noise and signal trials. Again using the assumption that the observer's responses on different trials are independent, the joint probability of x_f and x_h is the product of the individual probabilities:

$$P(x_f, x_h) = P(x_f; N_n, P_F)P(x_h; N_s, P_H).$$

This distribution is the product of two binomial distributions and is known as a *product-binomial distribution*. The same principles extend to more responses and to several stimuli. With r response alternatives, such as in a rating-scale study, the joint distribution of frequencies is multinomial (Equation A.37 on page 236):

$$P(x_1, x_2, \ldots, x_r) = \frac{N_i!}{x_1!x_2!\cdots x_r!} P_1^{x_1} P_2^{x_2} \cdots P_r^{x_r}.$$

The distribution of the complete set of data for t stimulus types is the product of t such terms, one for each stimulus. After rearranging its parts, the resulting *product-multinomial distribution* has the distribution function

$$P(\text{data}) = \left[\prod_{i=1}^{t} \frac{N_i!}{x_{i1}! x_{i2}! \cdots x_{ir}!} \right] \left[\prod_{i=1}^{t} \prod_{j=1}^{r} P_{ij}^{x_{ij}} \right]. \tag{11.1}$$

The product-multinomial sampling distribution, like any probabilistic model, is an idealization, and the two assumptions that underlie it (identical distribution and independence) are always violated to some degree. Depending on the application, these violations may or may not be serious. Before using any result that follows from these distributions, one should consider the actual process that generated the data and decide whether the violations are substantial enough to render their application dubious. These violations arise from the other sources of variability mentioned in Section 11.1.

Violations of identical distribution occur when the parameters of the model change during the collection of a set of data. This variation can be caused by either interobserver or intraobserver variability. Observers usually differ in their fundamental parameters and consequently in their probabilities of making a particular response to a stimulus. Their properties also can change both throughout a session and from one session to another. Practice and fatigue effects were mentioned above. Slightly less apparent are the changes that an efficient observer makes while adapting the decision criterion to the perceived proportion of signals (recall the discussion of the optimal criterion in Section 2.5). The effects of either of these sources of variability are similar. First, the parameters estimated from the composite data may not be those that describe the performance of any one observer or session (recall Section 4.6), and, second, the standard errors obtained from the multinomial distribution are underestimates of the real uncertainty.

Violations of independence occur when the events on one trial influence those on another trial. Usually these influences take the form of sequential dependencies, which can arise for several reasons. Instead of judging each stimulus in isolation, as implied by the model, the observer may judge it with reference to the previous trial or to the memory of that trial. Comparisons of this type would make a signal more likely to be detected when it follows a noise stimulus than when it follows another signal. Another source of dependence is the tendency of many observers to fall afoul of the *gambler's fallacy* by avoiding the response made on the previous trial. After saying YES three times in a row, such an observer is more likely to say NO, regardless of the stimulus.

The best cure for these problems is to use a well-practiced observer. In a well-run study, time should be allowed for the observer to become familiar with the task and to develop a reasonably stable pattern of behavior before starting to record data. If repeated sessions are used, some warm-up trials that are not recorded should be included at the start of each session, and the sessions themselves should be short enough that the observer does not become fatigued. Trained observers who understand the task well are less likely to show either parameter variability or sequential dependencies than are naive observers.

Violations of the product-multinomial assumptions, while almost inevitable, do not necessarily preclude using the statistical procedures described here. In many studies, the violations are small enough that a statistical treatment based on the product-multinomial distribution is still valuable. Even when the violations are substantial, they affect some parts of the procedures more than others. Estimates of proportions are least affected, and variances and standard errors are most affected. As a result, estimates of detectability and bias can be satisfactory, even though confidence intervals and statistical tests are questionable. Because almost all data are affected by sources of variability that are not accounted for by the multinomial distribution, they are overdispersed relative to it. Calculations based on it tend to underestimate the variability, so that confidence intervals are too short and tests are too likely to reject hypotheses.

11.3 Simple detection statistics

Although the product-multinomial distribution is fundamental to sampling theory in signal-detection theory, it is a means, not an end. The quantities of primary interest to a researcher are detection parameters such as d', A_z, or $\log \beta$. The standard errors of these parameters follow from the properties of the product-multinomial distribution. These standard errors let one know how much sampling effects are likely to make the parameter estimates fluctuate and how much one can rely on them. The standard errors then are used to construct confidence intervals and to run hypothesis tests.

The most commonly used statistical procedures are those that apply to simple detection. Data from these studies consist of two proportions, the hit rate h and the false-alarm rate f, and the various detection statistics are functions of these proportions:

$$\widehat{\lambda} = -Z(f),$$
$$\widehat{d'} = Z(h) - Z(f),$$

$$\widehat{\log \beta} = \tfrac{1}{2}[Z^2(f) - Z^2(h)],$$

$$\widehat{A_z} = \Phi(\widehat{d'}/\sqrt{2}) = \Phi[Z(h) - Z(f)]/\sqrt{2}.$$

(Equations 2.3, 2.4, 2.11, and 4.8, respectively). The uncertainty in these statistics derives from that of h and f, which, in turn, originates in their multinomial sampling distribution (Equation A.41 on page 237):

$$\widehat{\text{var}}(f) = \frac{f(1-f)}{N_n} \quad \text{and} \quad \widehat{\text{var}}(h) = \frac{h(1-h)}{N_s}. \qquad (11.2)$$

These sampling variabilities must be propagated to those of the detection statistics.

The conversion of the sampling variances of f and h to those of the detection statistics makes use of an important theorem in error propagation theory. It states that the variance of a transformation of a random variable is approximately proportional to the variance of the original variable, with the proportionality constant equal to the square of the derivative of the transformation.[1] If $Y = g(X)$, then

$$\sigma_Y^2 \approx \left(\frac{dg(x)}{dx}\right)^2 \sigma_X^2. \qquad (11.3)$$

This approximation is adequate when the samples are large and the standard errors are small.

All the detection statistics depend on the Gaussian transform of the original proportions. Applying Equation 11.3 to $Y = Z(P)$ gives

$$\sigma_Y^2 = \frac{\sigma_P^2}{\varphi^2(y)}.$$

The sampling variance of $\widehat{\lambda}$ is a direct application of this result to the false-alarm rate. Expressed as an estimate,

$$\widehat{\text{var}}(\widehat{\lambda}) = \frac{\widehat{\text{var}}(f)}{\varphi^2(\widehat{\lambda})}, \qquad (11.4)$$

with $\widehat{\text{var}}(f)$ given by Equation 11.2. The statistic d' is the difference of two independent terms, one from the false alarms and one from the hits. Its variance is the sum of the variance components (Equation A.33 on page 235):

$$\widehat{\text{var}}(\widehat{d'}) = \frac{\widehat{\text{var}}(f)}{\varphi^2(\widehat{\lambda})} + \frac{\widehat{\text{var}}(h)}{\varphi^2(\widehat{d'} - \widehat{\lambda})}. \qquad (11.5)$$

---

[1] This rule is obtained by approximating the nonlinear transformation $g(X)$ by a straight line near the observation point, then using the slope of this line to stretch or shrink the standard error.

This sampling variance is always greater than that of the criterion. Finding the sampling variance of the bias requires an additional application of Equation 11.3 to take care of the squares. It gives

$$\widehat{\text{var}}(\log \beta) = \frac{\hat{\lambda}^2 \widehat{\text{var}}(f)}{\varphi^2(\hat{\lambda})} + \frac{(\hat{d}' - \hat{\lambda})^2 \widehat{\text{var}}(h)}{\varphi^2(\hat{d}' - \hat{\lambda})}. \tag{11.6}$$

Finally, the sampling variance of the area under the operating characteristic is found by transforming the variance of d' from Equation 11.5, again using Equation 11.3:

$$\widehat{\text{var}}(\hat{A}_z) = \tfrac{1}{2}\varphi^2(\hat{d}'/\sqrt{2})\widehat{\text{var}}(\hat{d}'). \tag{11.7}$$

One should remember that these sampling variances are large-sample approximations. They are inaccurate when the number of trials is small, when h is near one, or when f is near zero.

Example 11.1: In the first session of the experiment described in Section 1.2 the false-alarm rate and the hit rate were $f = {}^{46}/_{100}$ and $h = {}^{82}/_{100}$. In Example 2.2 on page 24, the detection statistics were calculated to be $\hat{\lambda} = 0.100$, $\widehat{\log \beta} = -0.414$, $\hat{d}' = 1.016$, and $\hat{A}_z = 0.764$. Find the standard errors of these statistics.

Solution: To find the sampling errors, start by estimating the sampling variance of the false-alarm and the hit proportions (Equations 11.2):

$$\widehat{\text{var}}(f) = \frac{f(1-f)}{N_n} = \frac{0.46 \times 0.54}{100} = 0.002484,$$

$$\widehat{\text{var}}(h) = \frac{f(1-h)}{N_s} = \frac{0.82 \times 0.18}{100} = 0.001476.$$

The Gaussian densities of the noise and signal distributions at the criterion are

$$\varphi(\hat{\lambda}) = \frac{1}{\sqrt{2\pi}}e^{-\hat{\lambda}^2/2} = 0.3870 \quad \text{and} \quad \varphi(\hat{d}' - \hat{\lambda}) = 0.2622.$$

Using these values as intermediate results in Equations 11.4–11.7 gives

$$\widehat{\text{var}}(\hat{\lambda}) = \frac{\widehat{\text{var}}(f)}{\varphi^2(\hat{\lambda})} = \frac{0.002484}{(0.3870)^2} = 0.01576,$$

$$\widehat{\text{var}}(\hat{d}') = \frac{\widehat{\text{var}}(f)}{\varphi^2(\hat{\lambda})} + \frac{\widehat{\text{var}}(h)}{\varphi^2(\hat{d}' - \hat{\lambda})} = \frac{0.002484}{0.3870^2} + \frac{0.001476}{0.2622^2} = 0.03722,$$

$$\widehat{\mathrm{var}}(\widehat{\log \beta}) = \frac{\widehat{\lambda}^2 \widehat{\mathrm{var}}(f)}{\varphi^2(\widehat{\lambda})} + \frac{(\widehat{d'} - \widehat{\lambda})^2 \widehat{\mathrm{var}}(h)}{\varphi^2(\widehat{d'} - \widehat{\lambda})}$$

$$= \frac{(0.100)^2 \times 0.002484}{(0.3870)^2} + \frac{(0.916)^2 \times 0.001476}{(0.2622)^2} = 0.01817,$$

$$\widehat{\mathrm{var}}(\widehat{A_z}) = \tfrac{1}{2}\varphi^2(\widehat{d'}/\sqrt{2})\widehat{\mathrm{var}}(\widehat{d'}) = \tfrac{1}{2}(0.3082)^2 \times 0.03722 = 0.001768.$$

The standard errors of the statistics are the square roots of the variances:

$$\widehat{\mathrm{se}}(\widehat{\lambda}) = 0.126, \quad \widehat{\mathrm{se}}(\widehat{d'}) = 0.193, \quad \widehat{\mathrm{se}}(\widehat{\log \beta}) = 0.135, \quad \widehat{\mathrm{se}}(\widehat{A_z}) = 0.042.$$

To get a feeling for what standard errors to expect for the detection measures, it is helpful to consider an unbiased observer in an experiment in which signal and noise trials are equally frequent. This observer puts the criterion between the noise and signal distributions, optimally at $\lambda^\star = d'/2$ (Equation 2.16). With this criterion, the hit rate is one minus the false-alarm rate, and their sampling variances are the same:

$$\mathrm{var}(f) = \mathrm{var}(h) = \frac{\Phi(d'/2)\Phi(-d'/2)}{N}.$$

Substituting these values at the optimal criterion into Equation 11.5 gives

$$\widehat{\mathrm{var}}(\widehat{d'}) = \frac{2\Phi(d'/2)\Phi(-d'/2)}{N\varphi^2(d'/2)}. \tag{11.8}$$

This variance is the smallest that can be obtained for an observer with this d' and N.

The way that $\widehat{\mathrm{var}}(\widehat{d'})$ and $\widehat{\mathrm{var}}(\widehat{A_z})$ [obtained from $\widehat{\mathrm{var}}(\widehat{d'})$ using Equation 11.7] vary with d' is plotted in Figure 11.1 for a study with $N = 100$ trials of each type. The plot shows clearly how the estimate of d' becomes less reliable as the detectability increases. In contrast, because A_z is bounded by one, its variability is smaller at high values. This relationship lay behind the recommendation made in Section 4.6 to choose A_z over d' as a subject-level measure of detection when the detection rates are high.

A simple adjustment lets the plots in Figure 11.1 be used to roughly estimate the standard error for experiments with other than 100 trials per stimulus. The size of the standard error is proportional to $1/\sqrt{N}$, so values from the plot are corrected to those for a different N by multiplying by $10/\sqrt{N}$. For example, the standard errors are halved in a study with $N = 400$ trials per stimulus. Although this result applies only to an unbiased observer, and underestimates the true standard error for other observers, it gives a reasonable approximation unless the conditions are very asymmetrical.

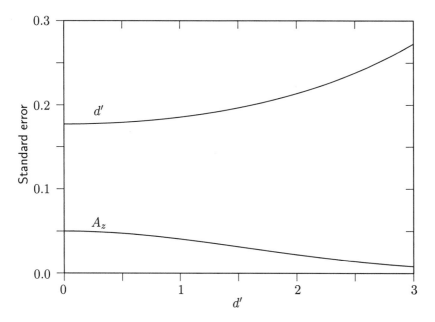

Figure 11.1: *Large-sample standard errors of estimates of d' and A_z as a function of the true detectability of the signal for an experiment with 100 noise trials and 100 signal trials.*

Example 11.2: A detection study is to be run with 250 trials of each type. About how accurate will an estimate of $\widehat{d'} = 1.4$ be?

Solution: From the plot at an abscissa value of 1.4, a standard error of about 0.195 is obtained. Then, correcting this value for the new sample size,

$$\widehat{se}(\widehat{d'}) \approx \frac{10}{\sqrt{250}} 0.195 = 0.12.$$

This value is an underestimate unless the bias is small.

The number of parameters in the equal-variance Gaussian model for simple detection equals the number of data points. This equality makes it possible to derive the sampling variances directly from the computing equations. That is not the case when the number of data points exceeds the number of parameters, as it does for most rating-scale data and for data from several bias conditions. Estimates then must be obtained using the fitting algorithms mentioned in Section 3.6. These algorithms can give approximate sampling variances of the parameters, based on the product-multinomial sampling models.

11.4 Confidence intervals and hypothesis tests

Two important inferential techniques, the construction of a confidence interval for a parameter and the test of a hypothesis about the parameters, are based on the standard error. In signal-detection theory, confidence intervals are most useful in describing the accuracy with which a parameter has been measured and the hypothesis tests are most useful in comparing two conditions.

The most direct use of the standard errors is to construct a *confidence interval* for a parameter. These intervals give a range of parameter values that are consistent with the outcome of a study. Thus, an interval $A \leq d' \leq B$ indicates that any value of d' between A and B is reasonably likely to have led to the data that were observed. Specifically, a $1 - \alpha$ confidence interval contains the set of parameter values for which the probability of the obtaining an observation more discrepant than the data is not less than α.

To construct a confidence interval, one needs information about the shape of the sampling distribution. When this distribution is approximately Gaussian, standard normal-distribution methods from statistics can be used. A $1-\alpha$ confidence interval lies between endpoints set $z_{\alpha/2}$ standard errors to either side of the observed value. The formula for a confidence interval for the mean, as presented in most elementary statistics texts and adapted to d', is

$$\widehat{d'} - z_{\alpha/2}\,\widehat{\text{se}}(\widehat{d'}) \leq d' \leq \widehat{d'} + z_{\alpha/2}\,\widehat{\text{se}}(\widehat{d'}). \tag{11.9}$$

This equation works for moderate values of d' and large samples, but with an important caveat. The standard errors used in Equation 11.9 should be those of the lower and upper endpoints, not that calculated for the observed value of d', which lies between them. Thus, the formula is correct only when the standard error does not change with the parameter value. In fact, as Figure 11.1 shows, neither d' nor A_z has a constant standard error. Because the standard error of $\widehat{d'}$ increases with d', an interval calculated under the assumption of homogeneous variance is somewhat too long in the lower end and somewhat too short in the upper end. The reverse is true for A_z.

In a correctly calculated interval, one must find a lower limit that is $z_{\alpha/2}$ standard errors away from d' using the standard error at that limit and an upper limit that is the same number of standard errors away from d' using the standard error at that limit. Finding these limits analytically requires solving an equation involving Equation 11.5, which can only be done numerically. A program to construct the interval numerically is easy to write. A good approximation to the size of the interval is obtained by adjusting Equation 11.9 using Equation 11.8 or Figure 11.1. Start with Equation 11.9, find the standard errors at the endpoints, and use them to recalculate the interval. For greater accuracy, the process can be repeated, but that is seldom necessary.

Example 11.3: Find a 90% confidence interval for the detection statistic d' in Example 11.1.

Solution: The data gave the estimate $\widehat{d'} = 1.016$ with $\widehat{se}(\widehat{d'}) = 0.193$. From the table on page 250, a 90% interval uses the critical value $z_{0.05} = 1.645$ that cuts off 5% of a Gaussian distribution in one tail. Using Equation 11.9 gives the interval

$$\widehat{d'} - z_{\alpha/2}\,\widehat{se}(\widehat{d'}) \le d' \le \widehat{d'} + z_{\alpha/2}\,\widehat{se}(\widehat{d'}),$$

$$1.016 - (1.654)(0.193) \le d' \le 1.016 + (1.654)(0.193),$$

$$0.70 \le d' \le 1.33.$$

This interval does not take the varying standard error into account. Looking at Figure 11.1, the range of the interval (0.7 to 1.3) falls in a region where the curve of the standard error does not rise greatly. So, although the interval should be a little shorter on the lower end and a little longer on the upper, it should not change substantially. Although the observer was not completely unbiased, Equation 11.8 (or Figure 11.1) can be used to estimate the size of the necessary adjustments. It reveals that the standard error at the calculated lower endpoint is about 3% shorter than in the center and that at the upper endpoint is about 4% longer. Making these adjustments gives an interval $0.71 \le d' \le 1.35$. The correction is negligible here, but would be larger if the interval were wider or were centered around a larger value of d'.

The asymmetry of the confidence interval is of less concern when the total width of the interval is of importance, not the particular endpoints. The width is important in planning a study, where it gives an indication of the precision to be expected in the results. It is roughly equal to $2z_{\alpha/2}\,se(d')$ for all but very large values of d'.

Example 11.4: A study is being planned in which three yes/no detection conditions will be run. These conditions are expected to yield d' statistics of about 0.6, 1.3, and 2.1. The values of d' are to be estimated with an uncertainty of no more than about ± 0.1, as determined by an 80% confidence interval. How many trials must be run in each condition?

Solution: Demanding an accuracy of ± 0.1 implies that the 80% confidence interval has a length of 0.2. The 80% confidence interval has limits at $\widehat{d}' \pm 1.28\,\widehat{\mathrm{se}}(\widehat{d}')$ (the Gaussian scores of ± 1.28 cut 10% in each tail), so has a width of $2.56\,\widehat{\mathrm{se}}(\widehat{d}')$. This width must equal the desired value of 0.2, which implies that $\widehat{\mathrm{se}}(\widehat{d}') = 0.2/2.56 = 0.078$. Because the standard error increases with d' (Figure 11.1), the condition in which the largest d' is expected has the most stringent requirement. Figure 11.1 shows that, had $N = 100$ been used with $d' = 2.1$, a standard error of about 0.22 would have resulted. To make the standard error equal to 0.078, it must be decreased by a factor of $0.22/0.078 = 2.82$. The sample size in the planned experiment, which goes up with the square this factor, must be $(2.82)^2 = 7.95$ times larger. A sample of about 750 to 800 trials per stimulus is needed. The large number of observations needed to put tight limits on d' is noteworthy.

When testing hypotheses about the parameter values, one does not need to worry about the problem of asymmetry. The standard errors in these tests are evaluated only at the hypothesized point. A test of the null hypothesis that a parameter takes a particular value is straightforward. Again using d' as an example, the null hypothesis that $d' = d'_0$ is tested by measuring the deviation between the observed and hypothesized value in units of the standard error. The test statistics is

$$z = \frac{\widehat{d}' - d'_0}{\widehat{\mathrm{se}}(\widehat{d}')}, \tag{11.10}$$

with the standard error evaluated at d'_0. Under the null hypothesis, z has approximately a standard Gaussian distribution with zero mean and unit variance, so the Gaussian tables on page 249 give critical values. For a two-sided test at the α level, one compares $|z|$ to the value $z_{\alpha/2}$ that cuts off the area $\alpha/2$ in each tail. If $|z| > z_{\alpha/2}$, then one rejects the hypothesis that $d' = d'_0$, and if $|z| \leq z_{\alpha/2}$, then the hypothesized value d'_0 remains tenable. Specifically, a test at the 5% level uses the familiar critical value of 1.96. Tests of hypotheses about other measures use the same procedure.

Example 11.5: In a recognition memory study, 45 of 60 old items were identified as OLD, as were 21 of 60 new items. Does this result indicate that the observer is remembering these materials?

Solution: If nothing were remembered, then $d' = 0$. The actual value observed was

$$\widehat{d}' = Z(h) - Z(f) = Z(0.75) - Z(0.35) = 1.06.$$

Could this value of $\widehat{d}'$ reasonably have been obtained were $d' = 0$ correct? The standard error of d' could be approximated from Figure 11.1, but it is better to let it accommodate any response bias that is present. If d' were zero, then the true probabilities of hits and false alarms would be identical. As there were $60 + 21 = 81$ OLD responses to the 120 items, an estimate of the probability of saying OLD is $p = 81/120 = 0.675$, corresponding to a criterion set at $\widehat{\lambda} = -Z(p) = 0.454$. Equation 11.5 now gives the standard error of d' estimated in a study with this bias. Under the null hypothesis, h and f have the same sampling distribution, and their common standard error (Equation 11.2) is

$$\widehat{\mathrm{var}}(f) = \widehat{\mathrm{var}}(h) = \widehat{\mathrm{var}}(p) = \frac{p(1-p)}{N} = \frac{0.675 \times 0.325}{60} = 0.00365.$$

Adapting Equation 11.5 to $d' = 0$ gives the standard error of d'

$$\widehat{\mathrm{var}}(\widehat{d'}) = 2\frac{\widehat{\mathrm{var}}(p)}{\varphi^2(\widehat{\lambda})} = \frac{2 \times 0.00365}{(0.442)^2} = 0.03737.$$

The test statistic for the hypothesis that $d' = 0$ (Equation 11.10) is

$$z = \frac{\widehat{d'} - d'_0}{\widehat{\mathrm{se}}(\widehat{d'})} = \frac{1.06 - 0}{\sqrt{0.03737}} = 5.48.$$

This value substantially exceeds the critical value of 1.96 for a test at the 5% level. Clearly, $d' > 0$, and the observer is remembering the material.

Exact hypotheses about parameter values (other than that they are zero) are rarely needed, and a more useful test is one that compares the values obtained from two independent conditions. The variance of the difference between two independent quantities is the sum of their individual variances. Under the same large-sample assumptions that are required for simple tests and confidence intervals, the hypothesis that the criterion is the same in two groups is tested by a z statistic similar to Equation 11.10,

$$z = \frac{\widehat{\lambda}_1 - \widehat{\lambda}_2}{\widehat{\mathrm{se}}(\widehat{\lambda}_1 - \widehat{\lambda}_2)} = \frac{\widehat{\lambda}_1 - \widehat{\lambda}_2}{\sqrt{\widehat{\mathrm{var}}(\widehat{\lambda}_1) + \widehat{\mathrm{var}}(\widehat{\lambda}_2)}}. \tag{11.11}$$

Other parameters, such as d' or A_z, are compared in the same way.

Example 11.6: It was argued in Example 2.2 on page 24 that the two sessions described in Section 1.2 do not differ in the detectability of the signals, but that they do in the criteria and biases used by the observer. Support these assertions.

Solution: Standard errors for the first session were calculated in Example 11.1. These results, with those for the second condition (also based on $N_n = N_s = 100$), are

		f	h	$\widehat{\lambda}$	$\widehat{d'}$	$\widehat{\log \beta}$
Session 1		0.460	0.820	0.100	1.016	−0.414
	S.E.			0.126	0.193	0.135
Session 2		0.190	0.055	0.878	1.004	0.377
	S.E.			0.144	0.192	0.128

Using these results, the standard error of the difference between λ_1 and λ_2 is

$$\widehat{se}(\widehat{\lambda}_1 - \widehat{\lambda}_2) = \sqrt{\widehat{var}(\widehat{\lambda}_1) + \widehat{var}(\widehat{\lambda}_2)} = \sqrt{0.126^2 + 0.144^2} = 0.191.$$

The test statistic comparing the two criteria is

$$z = \frac{\widehat{\lambda}_1 - \widehat{\lambda}_2}{\widehat{se}(\widehat{\lambda}_1 - \widehat{\lambda}_2)} = \frac{0.100 - 0.878}{0.191} = -4.07.$$

This statistic exceeds (in absolute value) the 5% critical value of 1.96. The hypothesis of equal criteria is easily rejected. The comparable standard errors for the sensitivity and the bias are $\widehat{se}(\widehat{d'}_1 - \widehat{d'}_2) = 0.272$ and $\widehat{se}(\widehat{\log \beta}_1 - \widehat{\log \beta}_2) = 0.186$, leading to test statistics of 0.04 and −4.25, respectively. There is no evidence to support a difference in the detectability, but the biases, like the criteria, unquestionably differ.

When using tests based on Equation 11.11, such as those in Example 11.6, one should remember that the standard errors come from large-sample approximations. If N is small or either f or h is near the limits of 0 and 1, then the standard errors are inaccurate and the tests are of limited accuracy. In such cases one should not depend on either hypothesis tests or confidence intervals. However, a glance at the calculated standard errors, although almost certainly optimistically too small, can still warn one away from overinterpreting small differences. This problem did not arise with the one-sample tests of the hypothesis $d' = 0$ or $A_z = 0$ (e.g., Example 11.5), where the standard error is based on the null value. Observations that differ greatly from the hypothesized value do not affect the test—they only indicate that the null hypothesis is very much wrong.

For the combination of the variances in Equation 11.11 to be correct, the two observations must be independent. There are several situations where one is tempted to apply this test, but where independence fails. One situation occurs when the observations are independent, but not the derived performance statistics. For example, suppose two signals are to be compared to see which is most detectable. To make the conditions most comparable, the comparison is made in a single experiment by presenting

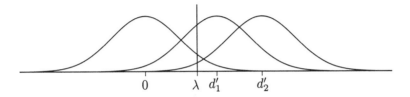

Figure 11.2: *The distributions of one noise and two signal stimuli.*

the observer with three types of stimulus: noise, signal S_1, and signal S_2. Two d' values are calculated, one based on the noise and S_1 results, the other on the noise and S_2 results. Two types of signal trial are used, but there is only one set of noise trials with which to compare them. Because the same noise distribution is used to calculate both values of d', they are not independent. Summing the estimated variances, as in Equation 11.11, overestimates the variability of the difference and gives the tests a negative bias.

A test for the difference between two signals in this circumstance is not hard to develop. Consider the underlying representation in Figure 11.2. There are two signal distributions and one noise distribution. A test of the hypothesis that $d'_1 = d'_2$ amounts to a test of whether the two signal distributions differ in location. For this comparison, the position of the noise distribution is irrelevant. An estimate of the difference between the d' statistics does not depend on the noise condition:

$$\widehat{d'_1} - \widehat{d'_2} = [Z(h_1) - Z(f)] - [Z(h_2) - Z(f)] = Z(h_1) - Z(h_2).$$

The proper sampling variance of this difference is the sum of the terms associated with the Z transformations of the two hit rates:

$$\widehat{\mathrm{var}}(\widehat{d'_1} - \widehat{d'_2}) = \frac{\widehat{\mathrm{var}}(h_1)}{\varphi^2(\widehat{d'_1} - \widehat{\lambda})} + \frac{\widehat{\mathrm{var}}(h_2)}{\varphi^2(\widehat{d'_2} - \widehat{\lambda})}.$$

The z statistic can then be used.

A second cause of dependent observations is more subtle. It occurs when independence fails because the observations themselves are dependent. For example, suppose the sensitivities of two observers are to be compared in a single experiment. A single series of trials is presented to both observers. Each observer gives a separate response on each trial. For each observer, a hit rate and a false alarm rate are found, and d' is calculated. This dependence is particularly marked when a large component of variability is intrinsic to the signals. Suppose the observers are detecting hidden elements in natural situations—camouflaged objects in pictures, tumors in

X-ray films, or voices amid a background of multiple conversations. For some instances the target object is very poorly hidden and both observers say YES, that is, both make a hit. For other instances the background is more confusing and both observers are likely to make errors, misses or false alarms, as the case may be. Because both observers see the same trials, their responses are not independent. This dependence invalidates the standard errors calculated by Equation 11.11.

This situation is more difficult to treat. The two observations on a trial must now be thought of as realizations of a bivariate variable, and the correlation between them must be estimated and used to find the standard error of the difference between the means. In effect, a single detection model must be fitted to both observers' data simultaneously. The techniques for fitting such a model and testing for equal detectability have been developed (see reference notes), but are beyond the scope of this book.

11.5 Goodness-of-fit tests

After fitting a theoretical model to a set of data, one often wants to know how well that model fits. The parameter estimates from a well-fitting model can be trusted, but those from a model that fits poorly can be deceptive. By looking at the quality of the fit, one can decide whether the model provides an adequate description of the data. A failure to fit often suggests that modification of the description is needed. For example, when an equal-variance Gaussian model fails to fit a set a data, one might want to relax the variance assumption or to change to a non-Gaussian model. Conversely, the satisfactory fit of a simple model saves one the cost, in data, time, and complexity, of working with a more complex description.

A test of a theoretical model is only possible when the data are sufficiently rich. For a test to be made, there must be ways in which the data could vary that cannot be accommodated by the model. A model that can agree with every possible configuration of observations cannot be tested. Specifically, there is no way to test whether the equal-variance Gaussian model is a correct description of a single yes/no detection study. There are two parameters in the model (d' and λ) and two independent pieces of data (h and f). It is always possible to choose the theoretical parameters to match the data exactly, regardless of the appropriateness of the model. As a general rule, for a model to be testable, it must have fewer parameters than there are independent pieces of data. The data must also provide information that is relevant to the aspect of the model that is being tested, which it does for all the models discussed in this book.

The basic goodness-of-fit testing strategy for signal-detection data follows the chi-square procedure described on page 245. The data form a table with entries x_{ij} in which the rows are stimulus type, the columns are responses, and the entries are the number of times that combination of stimulus and response occurred. Any theoretical model, once its parameters are estimated, implies a comparable table of predicted frequencies m_{ij}. If the two sets of frequencies are similar, then the model fits; if they differ substantially, then the model fails. Differences between the observed and fitted values are calculated with the Pearson statistic (Equation A.54; the likelihood-ratio statistic, Equation A.55, is less commonly used). When the degree of discrepancy is sufficiently great, doubt is thrown on the model.

The degree of freedom for these tests equals the number of cells less the number of constraints and the number of parameters (Equation A.56). When the observer can choose among r responses, there is generally one constraint (the total number of trials) for each stimulus type. If a p parameter model is fitted to data involving t stimulus types (commonly $t = 2$, for noise and signal) the number of degrees of freedom for a test is

$$\text{degrees of freedom} = t(r - 1) - p. \qquad (11.12)$$

Example 11.7: Use a goodness-of-fit test to establish that performance in the memory experiment of Example 11.5 is better than chance.

Solution: The test is made by creating a theoretical model of chance performance, then determining whether it fits. The observed responses x_{ij} form the table

	NO	YES	
New	39	21	60
Old	15	45	60
	54	66	120

Under a model of random performance, the responses are unrelated to the stimulus. The 66 YES responses would be, on average, evenly divided between the Old and New rows, as would be the 54 NO responses. The resulting table of expected frequencies m_{ij} is[2]

	NO	YES	
New	27	33	60
Old	27	33	60
	54	66	120

---

[2]The frequencies here are those of a chi-square test of independence, as presented in elementary statistics.

The Pearson X^2 statistic (Equation A.54) applied to these tables gives the value

$$X^2 = \sum \frac{(x_{ij} - m_{ij})^2}{m_{ij}}$$

$$= \frac{(39 - 27)^2}{27} + \frac{(21 - 33)^2}{33} + \frac{(15 - 27)^2}{27} + \frac{(45 - 33)^2}{33} = 19.39.$$

The model requires the fitting of one parameter, $P(\text{YES})$, so from Equation 11.12,

$$\text{degrees of freedom} = t(r - 1) - p = 2 \times (2 - 1) - 1 = 1.$$

This value is much larger than the chance value of 3.84 obtained from the chi-square table on page 251. The model of chance performance is rejected, and one concludes that the responses depend on the stimulus.

The null hypotheses tested in Examples 11.5 and 11.7 are the same, and it would be surprising if their results differed. In fact, both hypothesis were easily rejected. In the first example, $z = 5.48$ was obtained; in the second the X^2 statistic with one degree of freedom is equivalent to $z = \sqrt{19.39} = 4.40$. The difference between these values is not of substantive importance. It reflects the different modes of testing, which are sensitive to different forms of violation of the null hypothesis. The two procedures typically lead to the same decision.

Example 11.8: Test the goodness of fit of the equal-variance and the unequal-variance Gaussian models to the rating data in Table 5.1 on page 84, shown again in the upper block of Table 11.1.

Solution: First the models are fitted. The estimates made in Chapter 5 for the unequal-variance model had signal parameters of $\hat{\mu}_s = 1.325$ and $\hat{\sigma}_s = 1.360$, with criteria at $-0.714, -0.067, 0.423, 0.979$, and 1.590. For the equal-variance model, $\hat{d'} = 1.102$ with, by definition, $\sigma_s = 1$, and the criteria are $-0.623, -0.027, 0.409, 0.883$, and 1.371. The predicted probabilities for this model are obtained by inserting these parameters into Equations 5.3 and 5.4, and the expected frequencies are found by multiplying each probability by the number of trials of that type. For example, the expected number of Y2 responses to signals under the unequal-variance model is

$$P_{25} = \Phi\left(\frac{\hat{\lambda}_5 - \hat{\mu}'_s}{\hat{\sigma}_s}\right) - \Phi\left(\frac{\hat{\lambda}_4 - \hat{\mu}'_s}{\hat{\sigma}_s}\right)$$

	N3	N2	N1	Y1	Y2	Y3
	Original frequencies x_{ij}					
Noise	166	161	138	128	63	43
Signal	47	65	66	92	136	294
	Fitted frequencies m_{ij}: equal-variance model.					
Noise	186.34	155.60	118.58	106.58	72.39	59.51
Signal	29.59	61.04	80.38	118.29	135.07	275.64
	Fitted frequencies m_{ij}: unequal-variance model.					
Noise	166.06	164.88	133.09	120.55	75.30	39.11
Signal	46.91	60.36	70.31	102.21	124.11	296.10

Table 11.1: *Observed and fitted frequencies for the confidence-rating data in Table 5.1.*

$$= \Phi\left(\frac{1.590 - 1.325}{1.360}\right) - \Phi\left(\frac{0.979 - 1.325}{1.360}\right) = 0.1773,$$

$$m_{25} = N_s P_{25} = 700 \times 0.1773 = 124.11.$$

The complete sets of expected frequencies are given in the two lower blocks of Table 11.1. Examination of these frequencies suggests that the unequal-variance model fits well, but that the equal-variance model deviates systematically. The goodness-of-fit statistics support these observations. The Pearson statistic is $X^2 = 5.93$ for the unequal-variance model and $X^2 = 35.85$ for the equal-variance model. For both tests, $t = 2$ and $r = 6$, but the number of parameters in the models differs. There are $p = 6$ parameters for the equal-variance model (d' and five criteria), so

$$\text{degrees of freedom} = t(r - 1) - p = 2 \times (6 - 1) - 6 = 4.$$

From the table on page 251 the critical value at the 5% level for a chi-square distribution with 4 degrees of freedom is 9.49. The observed value is considerably greater, so the equal-variance model is rejected. The unequal-variance model has one more parameter, so one fewer degree of freedom. The critical value is 7.81, so the observed deviations are not sufficient to reject the model. The Gaussian representation is satisfactory.

Although the results in Example 11.8 are clear—the unequal-variance model is unambiguously preferable—the two goodness-of-fit tests do not directly compare the models. To make comparative assertions, it is necessary to directly pit one model against the other. Tests of this type are discussed in the following section.

Some attention should be paid to the assumptions of the goodness-of-fit tests. For proper reference to the chi-square distribution, three conditions are necessary. Two of these are the assumptions that give the data a product-multinomial distribution: the observations must be mutually independent and identically distributed. As discussed in Section 11.2, these assumptions are reasonably well satisfied when practiced observers and large samples are used, but may be violated with untrained or inexperienced observers. Pooling data across observers with different properties also violates the identical-distribution assumption.

The third condition concerns the number of observations. Because the multinomial distribution is discrete, the distribution of X^2 also is discrete. The continuous chi-square distribution only approximates its true form, and this approximation improves with the size of the sample. Conventionally, the approximation is considered to be satisfactory when the predicted frequency of every categorized event exceeds five. A few smaller cells are tolerable, particularly in larger tables and when the test has several degrees of freedom. In rating-scale experiments, the rareness of high-confidence YES responses to noise and NO responses to signals does not greatly affect the quality of the tests. However, one should distrust the goodness-of-fit tests when there are many expected frequencies that are much smaller than four or five, particularly when the observed statistic is close to the critical value.

The use of the statistical test here is somewhat different from its use in conventional null-hypothesis testing. In much research, one sets up a null hypothesis that is the opposite of what one is hoping to find and tries to show that it is false. For example, to establish that two groups have different means, one posits that they are equal, then shows that the data are inconsistent with this hypothesis. In such cases, a large test statistic value—a "significant" result—is what one wants to find. The model-fitting strategy differs from this approach in that one is interested in establishing a satisfactory model, not in rejecting it. One wants to find a small test statistic, indicating that the data do not substantially deviate from the model. Of course, one can never prove that a model is correct (nor is any model completely true in all its details). Even a perfect fit to data does not ensure that the processes embodied in the model are a correct description of how the observer operates. However, as long as the samples are large enough to give the test reasonable power, the lack of a significant value of the test statistic is an indication that the parameter estimates are a satisfactory representation of detection performance, within the context of the model.

Although the goodness-of-fit tests let one evaluate the appropriateness of a representation, their importance should not be overstated. The test statistics are more sensitive to violations of the assumptions than is the overall model. The parameters of the signal-detection models are useful summaries of detection or discrimination data, even when, strictly speaking, the models that underlie them can be rejected. In a large experiment, where many observations have been collected, the goodness-of-fit tests have very large power, and they can lead the model to be rejected for reasons that, while valid, are not sufficiently great to compromise the summary statistics. Common sense is needed here. When a model appears to fail, one should look at the data to see why it failed and decide what to do. Sometimes there is an obvious extension or modification that can improve the situation. Thus, for the rejection of the equal-variance model in Example 11.8, a quick look at the operating characteristic (Figure 5.2 on page 89) shows that the failure is an important one and that a different representation is necessary. Other times there are no systematic deviations, just scattered variation that is more or less uninterpretable. Here it is usually better to continue to use the derived detection statistics than to give up the analysis. Had X^2 for the unequal-variance model been about eight in Example 11.8, the model would strictly have been rejected, but it would still be appropriate to use statistics like A_z to describe the performance.

11.6 Comparison of hierarchical models

Although goodness-of-fit tests are very useful, they are inadequate for certain types of questions. It is common to find situations in which two representations, one a special case of the other, are to be compared. The prime example in signal-detection theory is of the equal-variance and unequal-variance Gaussian detection models. Both representations have in common the distributions $X_n \sim \mathcal{N}(0,1)$ and $X_s \sim \mathcal{N}(\mu_s, \sigma_s^2)$. The equal-variance representation is a special case of this general representation, obtained by fixing $\sigma_s^2 = 1$. These two descriptions constitute a hierarchical pair of models, one of which includes the other. One needs a way to decide whether the general (unequal-variance) version is a substantial improvement over the specific (equal-variance) version.

The goodness-of-fit tests cannot answer this question. The general model, being more flexible and including the restricted model as a special case, cannot fit the data less well than does the restricted model. Because the general model has greater flexibility, the statistics that measure its goodness of fit have fewer degrees of freedom than do those for the restricted model. The criterion used to evaluate the general model is more stringent. Thus, in Example 11.8 the critical value to reject the unequal-variance model was 7.81 and that for the equal-variance model was 9.49. With two separate test statistics having different degrees of freedom, it is possible for either description to be rejected or retained. Because null-hypothesis testing only assesses evidence and is not a exact logic, inconsistencies can arise. For example, had $X^2 = 8$ for the unequal-variance model and $X^2 = 9$ for the equal-variance model in Example 11.8, then the goodness-of-fit tests would have rejected the former model and retained the latter, notwithstanding the fact that it is inconsistent to claim that a description is false while a special case of it is true. A test specifically directed at the difference between the models is necessary to ensure consistent results.

Fortunately, such a test is easy to derive from the goodness-of-fit tests. Two hierarchical models can be compared by looking at the difference in their goodness-of-fit statistics. The theoretical basis for the comparison is better when one uses the likelihood-ratio statistic G^2 than when one uses the Pearson statistic X^2, although in most cases they yield the same conclusion. Formally, let $\mathcal{M}$ and $\mathcal{M}^\star$ denote two models for a set of data, with model $\mathcal{M}^\star$ being a special case of model $\mathcal{M}$. Each model is fitted and tested using Equation A.55, giving statistics $G^2(\mathcal{M})$ and $G^2(\mathcal{M}^\star)$, respectively. Because the general model contains more parameters than the restricted model, its test statistic has fewer degrees of freedom. The null hypothesis that the general model is no improvement over the restricted model is tested by the difference in fit:

$$G^2(\mathcal{M}^\star : \mathcal{M}) = G^2(\mathcal{M}^\star) - G^2(\mathcal{M}) \qquad (11.13)$$

with degrees of freedom

$$df(\mathcal{M}^\star : \mathcal{M}) = df(\mathcal{M}^\star) - df(\mathcal{M}). \qquad (11.14)$$

Substitution of Equation A.55 into Equation 11.13 shows that this difference test can be written directly as[3]

$$G^2(\mathcal{M}^\star : \mathcal{M}) = 2 \sum_i \sum_j x_{ij} \log \frac{m_{ij}}{m_{ij}^\star}. \qquad (11.15)$$

Example 11.9: Examine the appropriateness of the assumption of equal variance in the Gaussian models for the rating data in Table 5.1 and Example 11.8.

---

[3]This test is an instance of the likelihood-ratio testing procedure described in Section 9.1. When the product-multinomial probabilities from Equation 11.1 are formed into a log likelihood ratio $\log[\mathcal{R}(H_1 : H_2)]$ (Equation 9.8 on page 156) with $\mathcal{M}$ as H_1 and $\mathcal{M}^\star$ as H_2, Equation 11.15 is obtained, except for a factor of 2 needed to bring it in line with a chi-square distribution.

Solution: Expected frequencies under the two models were given in Table 11.1 on page 215. The likelihood-ratio goodness-of-fit statistics calculated separately for the two models (Equation A.55) are $G^2(\text{equal}) = 35.11$ and $G^2(\text{unequal}) = 6.02$. The difference between these values is

$$G^2(\text{equal : unequal}) = 35.11 - 6.02 = 29.09.$$

This statistic is compared to a chi-square distribution with $4 - 3 = 1$ degree of freedom. Its value exceeds any reasonable criterion based on this distribution, so the hypothesis of equal variance is rejected. The difference between the X^2 goodness-of-fit statistics gives a similar, albeit approximate, test statistic of $35.85 - 5.93 = 29.92$. Putting the observed and expected frequencies from Table 11.1 directly into Equation 11.15 gives the same result:

$$G^2(\text{equal : unequal}) = 166 \log \frac{166.06}{186.34} + 161 \log \frac{164.88}{155.60} + \cdots$$
$$+ 294 \log \frac{296.10}{275.64}$$
$$= -19.127 + 9.327 + \cdots + 21.051 = 29.09.$$

The hierarchical tests are very specific. All their power is focused on the parameter or parameters that differentiate the models. Thus, in Example 11.9, the equal-variance model failed to fit with $X^2 = 35.85$, but this statistic does not identify the cause of the problem. However, knowing that relaxing the equal-variance assumption leads to a large change in G^2 pinpoints the variance as the difficulty, at least in the context of the Gaussian model.

Another characteristic of the difference tests is that they have somewhat weaker large-sample requirements than do the goodness-of-fit tests. The use of the chi-square reference distribution for a goodness-of-fit test required a minimum expected frequency in each cell. In contrast, the chi-square approximation for the difference tests is less affected by small expected frequencies in individual cells, as long as the total number of observations is large. The asymptotic properties of the tests depend on the sample size per difference degree of freedom, not per cell. The difference tests can be satisfactory in sets of data for which the overall goodness-of-fit tests are doubtful or invalid.

The difference test involves one important condition: the more general model must fit the data adequately. If it is substantially wrong, then its parameters may have little meaning, and tests of their values are not interpretable. Poor fit is not a problem in Example 11.9, where the good fit of the unequal-variance model implies that the test of variance equality can be relied upon. Goodness-of-fit tests of the general model are helpful here, but one should not be overly fussy about them. Because of their sample-size requirements and the generality of their alternative hypotheses, they are usually less robust than the difference tests. The key requirement is that the general model does not grossly misrepresent the data.

A sophisticated analysis makes use of both difference and goodness-of-fit tests, the first to compare models and the second to see whether they are adequate. When several parameters are involved, it is helpful to think of a hierarchy of models, ranging from the most simple description through to a *saturated model* that perfectly agrees with the data. Each model is associated with its degree of goodness of fit, and each hierarchical connection between models is associated with a specific assumption about the parameters. Consider a study in which rating-scale responses are collected for the detection of two different stimuli. A complete Gaussian model for the study depends on two means, two variances, and two sets of criteria. One can then form a series of restricted models in which various parameters are fixed or equated across groups. Setting the two sets of criteria to be the same creates a test of whether the category ratings are used consistently, setting the signal variances to one creates a test of the equal variance assumption, and equating the two signal means creates a test of equal discriminability of the two stimuli (as measured by Δm). These restrictions could be applied either separately or in sequence. A researcher considering (either implicitly or explicitly) such a collection of hierarchical models can select the combination of difference tests and goodness-of-fit tests that best treats the results of a study.

To use the difference tests, the two models that are being compared must be hierarchically related. Models that do not have a hierarchical connection cannot be compared by these tests. Most important questions about the parameters of a class of models can be formulated as hierarchical tests, but nonhierarchical hypotheses arise when different representations are compared. When two competing models have the same complexity, one can go with the one that has the better fit, although this choice does not have a probabilistic basis. The situation is more problematic when the models differ in complexity. When a pair of nonhierarchical models have different numbers of parameters, even the better fit of the more complex model does not necessary imply that it is a better description of the underlying process. The good fit may be due only to its greater flexibility. Several schemes have been proposed to adjust the goodness-of-fit statistics for the number of parameters, but these lack a probabilistic basis.

These problems are illustrated by Example 8.2 on page 139, which compared the fits of the high-threshold model and a Gaussian model. These two models are not a hierarchical pair—they are conceptually unlike, and neither model reduces to the other. Each model can be fitted to the data, and their properties compared qualitatively by looking at the operating characteristics in Figure 8.5. However, the difference in their goodness-of-fit statistics is not meaningful. Of the two, the Gaussian model fits slightly better, but to get this improved fit it was necessary to go to the unequal-variance version (the equal-variance model clearly fails). That model has one more parameter than the high-threshold model. Whether the improved fit is worth the extra parameter depends more on other characteristics of the experiment than on any statistical index.

11.7   Interobserver variability

In the traditional signal-detection experiment, one well-practiced observer provides many observations in each condition. Any statistical treatment is based on multinomial sampling principles, and the reliability of a finding is established by replicating it with other observers. However, as was pointed out in Section 4.6, studies of this type are not always possible, nor are they consistent with the type of design that is necessary or deemed appropriate in many domains. Much research is conducted with larger groups of subjects, with each subject contributing only a small portion of the total data set. Most memory and cognitive research is of this type.

In some studies it is impossible to get measures of performance in several conditions from a single observer. Consider a study in which two different methods of training an observer to detect a signal are to be compared. The two methods cannot be measured in a single observer, as would be possible were the detectability of two lights of two different intensities being studied. Each individual can be trained in only one way, so separate observers must be used for the different conditions. However, the use of separate observers introduces the interobserver sources of variability. Had just two observers been used, one for each method, it would be impossible to tell whether any differences in performance were caused by the training or by differences in the observers' sensitivities that were present before the training started. To separate the effects, one must record data from two groups of subjects, and the study must address the difference between the groups.

The problem of aggregating detection statistics was discussed in Section 4.6, where it was pointed out that an estimate of the mean of d' over subjects (i.e., $\mu_{d'}$) is better found by averaging individual d' values than by taking a composite. Using the standard errors to get confidence intervals or to run hypothesis tests introduces some further issues related to variability. The presence of interobserver sources of variability has two important consequences, one involving the hypotheses to be tested, the other concerning the sources of variability against which these hypotheses are tested. In a single-observer experiment, the null hypothesis that is tested is that the detection levels in two (or more) conditions are the same. A typical null hypothesis is that $d'_1 = d'_2$. In a study with many subjects, the statistical inference must apply to the population from which the subjects were sampled, not to the individual subject or group of subjects. Any null hypothesis asserting equal values of d' is trivially false because these statistics differ among individuals even within a group. Instead, the null hypothesis must be expressed in terms of the mean value over the population of individuals, that is, as $\mu_{d'_1} = \mu_{d'_2}$.

Along with the change in null hypothesis, there is a change in the structure of variability in the problem. Because the subjects in the two groups are viewed as sampled from a larger population, the variability of a score combines both within-subject and between-subject components. The within-subject multinomial sampling variability, which was used to derive the estimates $\widehat{var}(f)$ and $\widehat{var}(h)$ in Equations 11.2, underestimates the true variability in the experiment.[4] Confidence intervals for the mean $\mu_{d'}$ that are based on these estimates are too short because they neglect the extent to which subjects vary. Perhaps more seriously, hypothesis tests, such as those comparing the training methods, are biased toward rejection of their null hypotheses. The bias is particularly severe when the responses are pooled over subjects before the signal-detection statistics are calculated. Variabilities calculated from multinomial sampling distributions do not reflect the subject differences.

The bias in the pooled estimate is apparent in Example 4.1 on page 80, which aggregated recognition performance from five subjects. The usual between-subjects estimate of the standard deviation of d', based on its $n = 5$ observed values, is

$$s_{\widehat{d'}} = \sqrt{\frac{\sum(\widehat{d'}_1 - \overline{d'})^2}{n-1}} = 0.538.$$

In contrast, applying the within-observer formula for the standard error of d' (Equation 11.5) to the average false-alarm and hit rates $\overline{f} = 0.560$ and $\overline{h} = 0.744$ gives an apparent standard error of 0.165. The latter value grievously overestimates the stability of the observations.

Statistical analysis in a between-subject design is both simple and familiar. The observed difference between the mean detection statistics in the two groups must be assessed relative to both the individual multinomial variability and the intersubject variability. The variability of the individual

---

[4]In the statistical literature, such data are said to be *overdispersed* relative to the multinomial model.

subjects' detection rates combines both of these sources, and the within-group variance measures both simultaneously. One can use an ordinary t test to make the comparison. Specifically, the detection level (or whatever other characteristic one wishes) is measured for each subject. These numbers are entered in the t formula familiar from elementary statistics. For designs that involve three or more groups, the analysis of variance is used.

When the amount of data from each subject is not great, some of the individual statistics may be indeterminate due to observed frequencies of zero. As mentioned at the end of Section 2.3, one should be careful about replacing the zero by some other small value because this arbitrary choice can influence the variance. A better solution is to use a statistical procedure that is not sensitive to specific numerical values, particularly at the extremes. One possibility is to use a test based on ranks, such as the *Mann-Whitney U test*. In this type of test, the rank order of the scores is treated as useful data, but not their particular numerical values. Scores that are indeterminately large or small are treated as ranking above or below the numerical scores, but not otherwise specified. Such tests are almost as powerful as the standard "parametric" tests, and they can be more so when ambiguous variability of the extreme scores is present.

In one respect, the between-subject comparisons are simpler than the single-subject analysis. Although measures of sampling variability based on the multinomial distribution can no longer be used, they are replaced with empirical estimates. These estimates are less dependent on sampling assumptions than were the multinomial-based measures. They do not require one to assume that the trials are independent or are identically distributed, as described at the end of Section 11.5. These distributional concerns are transferred from the trials to the subjects. As in any such inferential analysis, one must assume that the subjects' scores are independent of each other and that any groups are drawn from homogeneous and equivalent populations. It is far easier to satisfy these assumptions with good experimental practice than to ensure that the multinomial assumptions are satisfied.

Reference notes

The methods in this chapter are largely part of the standard statistical analysis of categorical data, much of which is based on product-multinomial sampling models. Any of several books on the subject can be consulted for more details (e.g., Agresti, 1990; Wickens, 1989). The goodness-of-fit tests are described in most elementary statistics texts. For the extension of the difference tests to correlated statistics, see Metz, Want, and Kronman (1984) or Wickens (1991). The tests based on between-observer variability that are discussed in Section 11.7 are described in most elementary or intermediate texts.

The relationship of multinomial variability to between-subject variability in categorical models is not completely worked out in the statistical literature. Some of the ideas presented here draw on my discussion in Wickens (1993).

Exercises

11.1. Find standard errors for the statistics $\widehat{d'}$, $\widehat{A_z}$, and $\widehat{\log \beta}$ calculated from the data used in Problem 2.6 and 4.2.

11.2. Calculate 90% confidence intervals for d' and A_z for Problem 11.1.

11.3. Find 95% confidence intervals for the area A_z and the bias $\log \beta$ in Problem 7.1.

11.4. Starting with the estimate of λ_{center} in Equation 2.6 on page 28, use the logic that led to Equations 11.4–11.7 to find a formula for its standard error. Apply your equation to Problem 11.1.

11.5. A program that fits the rating-scale model to the data of Problem 5.1 reports $\widehat{d'} = 1.161$ with a standard error of 0.163. Give a 95% confidence interval for this value using the symmetrical formula.

11.6. Two observers in a detection task (with different sequences of stimuli) gave the following results:

	Observer A				Observer B	
	NO	YES	and		NO	YES
Noise	127	23		Noise	125	25
Signal	48	102		Signal	34	116

Run appropriate tests to decide whether the observers differ in sensitivity and in bias.

11.7. Data from a hard discrimination task were

	A	B
Signal A	54	66
Signal B	38	82

Can you reject the hypothesis that the observer was unable to discriminate the stimuli?

a. Use a z statistic.

b. Use a goodness-of-fit test.

11.8. Which of the following models form hierarchical pairs where differences in fit can be tested by the likelihood-ratio statistic G^2? Indicate the degrees of freedom for these tests.

- The high-threshold model
- The low-threshold model
- The equal-variance Gaussian model
- The rating-scale model
- The unequal-variance Gaussian model
- The three-state model

11.9. The detection task in Problem 11.6 was repeated four eight days, giving the results

Day	|	Observer A					Observer B		
	|	C.R.	F.A.	Miss	Hit	C.R.	F.A.	Miss	Hit
1	|	48	102	127	25	34	116	125	25
2	|	126	24	43	107	127	23	33	117
3	|	116	34	48	102	130	20	27	123
4	|	128	22	37	113	126	24	36	114
5	|	116	34	32	118	125	25	23	127
6	|	127	13	41	109	126	24	28	122
7	|	125	25	42	108	130	20	31	119
8	|	121	29	44	106	130	20	32	118

Answer the same questions as in Problem 11.6, using the variability among days as a measure of error.

11.10. Conduct a statistical test appropriate to Problem 6.5.

Appendix

A summary of probability theory

A basic knowledge of probability theory is necessary for an understanding of signal-detection theory. Although most introductions to statistics include some coverage of probability theory, it is frequently insufficient for the material in this book. Hence, this appendix reviews the ideas and concepts from probability theory that are most relevant to signal-detection theory, and it introduces the notation that is used here. The coverage is necessarily limited and omits many of the technical and mathematical details.

A.1 Basic definitions

To begin, think of some process, observation, or experiment that yields one outcome from among a collection of possible outcomes. These outcomes are called *elementary events*, and the set of them is known as the *sample space* over which probability is defined. Sets of elementary events are called simply *events*. These events are indicated by upper.case letters, such as A, B, and so forth.

Because the events are sets of outcomes, the methods of set theory are used to combine them. In particular, the *union* of two sets, which is denoted by $A \cup B$, is the set of elements that is in A or B or both of them, and the *intersection* of the events, which is denoted by $A \cap B$, is the set of elements that is in both, also called a *joint event*. As a mnemonic to distinguish these symbols, remember that the "cup" symbol $\cup$, having an open top, can hold more than the "cap" symbol $\cap$, and that the union $A \cup B$ is larger than the intersection $A \cap B$. Two sets that contain no elements in common are *mutually exclusive*. For them, $A \cap B = \emptyset$, where $\emptyset$ is the *empty set* that contains no members.

The probability of an event A is a number between 0 and 1, symbolized by $P(A)$. When $P(A) = 1$, the event is almost sure to happen, and when $P(A) = 0$, it almost never happens—the "almost" is necessary to accommodate certain infinite sample spaces. Events with larger probabilities are more likely to happen than those with smaller probabilities. The rules for probability do not in themselves determine how to assign numbers to events. Any assignment of numbers is satisfactory, as long as it satisfies a few crucial properties. Frequently the values are based on some theoretical description or model.

A properly defined assignment of probabilities must obey several rules. These rules are the same as those that describe the areas of regions on a plain, so that probabilities can be represented pictorially by overlapping circles or, as will be more important here, by areas under curves. Two rules set the limits of the measure:

$$P(\emptyset) = 0 \quad \text{and} \quad P(\text{sample space}) = 1. \tag{A.1}$$

A third rule determines the probability that the event that A does not occur:

$$P(\text{not } A) = 1 - P(A). \tag{A.2}$$

The most useful rules concern combinations of events. For any two events, the probability of their union is

$$P(A \cup B) = P(A) + P(B) - P(A \cap B). \tag{A.3}$$

When the two events are mutually exclusive, their intersection is empty and the union is the sum of the individual probabilities:

$$P(A \cup B) = P(A) + P(B). \tag{A.4}$$

Mutually exclusive sets of events are important. A set of mutually exclusive events that exhaust the sample space is known as a *partition*. One and only one of these events is certain to happen. If the events B_1, B_2, ... , B_n are such a partition, then the sum of their probabilities is one. Moreover, the probability of any other event A is the sum of the probabilities of its intersection with the members of the partition:

$$P(A) = P(A \cap B_1) + P(A \cap B_2) + \cdots + P(A \cap B_n). \tag{A.5}$$

This relationship, known as the *law of total probability*, is very widely used.

It is sometimes more convenient to specify the probability of an event by its *odds*, which is the ratio of the probability that an event happens to the probability that it does not:

$$\text{odds}(A) = \frac{P(A)}{P(\text{not } A)}. \tag{A.6}$$

Thus, probabilities of $\frac{1}{4}$, $\frac{1}{2}$, and $\frac{2}{3}$ give odds of $\frac{1}{3}$, 1, and 2, respectively. Often it is more convenient to work with the natural logarithm of the odds (see note 3 on page 29), known as the *logit*:

$$\text{logit } A = \log \text{odds}(A) = \log \frac{P(A)}{P(\text{not } A)}. \tag{A.7}$$

For the three examples above, the logits are -1.099, 0, and 0.693. Among the advantages of the logit is the fact that its values are symmetrically placed around zero. For example, $\text{logit}(\frac{1}{4}) = -1.099$ and $\text{logit}(\frac{3}{4}) = +1.099$.

Conditional probabilities

The way in which one event influences another is embodied in the concept of *conditional probability*. The conditional probability of event A given event B is like the ordinary probability of A, except restricted to the cases in which B also occurs. It is calculated as the ratio of the joint probability of both events to the probability of the conditioning event:

$$P(A|B) = \frac{P(A \cap B)}{P(B)} \quad \text{and} \quad P(B|A) = \frac{P(A \cap B)}{P(A)}. \tag{A.8}$$

By rearranging these definitions, the joint probabilities can be calculated from the conditionals:

$$P(A \cap B) = P(A|B)P(B) = P(B|A)P(A). \tag{A.9}$$

Combining this equation with Equation A.5 gives another version of the law of total probability:

$$P(A) = P(A|B_1)P(B_1) + P(A|B_2)P(B_2) + \cdots + P(A|B_n)P(B_n). \tag{A.10}$$

An important use of Equation A.9 is to reverse the events in a conditional probability. By solving the right-hand part of this equation for $P(A|B)$, one obtains

$$P(A|B) = \frac{P(B|A)P(A)}{P(B)}. \tag{A.11}$$

This result is known as *Bayes' theorem*. Note that in it $P(A|B)$ appears on the left and $P(B|A)$ appears on the right.

Bayes' theorem is particularly important as a way to accommodate evidence. Suppose that A corresponds to a statement about the world and that you have some *prior probability* $P(A)$ that A is true. You then observe an event B that is probabilistically related to A, in the sense that $P(B|A) \neq P(B|\text{not } A)$. Bayes' theorem allows you to incorporate the information you obtain from B into your assessment of A. Combining Equation A.11 with Equation A.10, you determine the *posterior probability* of A:

$$P(A|B) = \frac{P(B|A)P(A)}{P(B|A)P(A) + P(B|\text{not } A)P(\text{not } A)}. \qquad (A.12)$$

Here $P(A)$ gives your initial assessment of the probability of A and $P(A|B)$ gives it after you know that B occurred. When $P(B|A)$ and $P(B|\text{not } A)$ are very different, the change can be large.

A conditional probability expresses how the probability of one event is affected by the occurrence of another, and one possibility is that the two events have no influence on each other. Two events are *independent* when the probability of either event is not altered by the occurrence of the other:

$$P(A|B) = P(A) \qquad \text{and} \qquad P(B|A) = P(B). \qquad (A.13)$$

These relationships are equivalent: one cannot hold without the other. Calculating the joint probability is easy when the events are independent. No conditional probabilities are necessary:

$$P(A \cap B) = P(A)P(B) \qquad (A.14)$$

(cf. Equation A.9). The probability of the union of two independent events can also be found from the individual probabilities; Equation A.3 becomes

$$P(A \cup B) = P(A) + P(B) - P(A)P(B). \qquad (A.15)$$

Equations A.14 and A.15 greatly simplify calculation with probabilities, and for this reason, it is worth some effort to construct models in terms of independent events.

A.2   Random variables

Much of probability theory deals with numerical outcomes. A *random variable* is used to express an uncertain numerical outcome. Each time that a random variable is observed, it may (but does not have to) take a different value. Throughout this book, random variables are denoted by uppercase letters, usually X, Y, or the like. The probability rule that governs the values that a random variable may take is known as its *distribution*.

There are several classes of random variable, with different properties and different ways to specify their distribution. The most important distinction is between a *discrete random variable*, whose values come only from a discrete set of possibilities, such as the positive integers, and a *continuous random variable*, whose values can take any real number in a finite or infinite range.

A discrete random variable is described by its *probability function*, which gives, either numerically or by a formula, the probability of particular outcomes. For example, a random variable that takes three values might have the distribution

$$P(X{=}1) = \tfrac{1}{2}, \qquad P(X{=}2) = \tfrac{1}{3}, \qquad \text{and} \qquad P(X{=}3) = \tfrac{1}{6},$$

while one that takes values on the positive integers might have probabilities determined by the rule $P(X{=}k) = e^{-k}$. In each case, specific probabilities are assigned to each value of X, and these values sum to one.

Continuous random variables, on the other hand, cannot be defined by probability functions. In any interval for X, however small, there are an infinite number of different values, and each of these cannot be given a positive probability without the whole thing adding up to more than one. So the probability of any absolutely exact value of X must be vanishingly small; $P(X = x)$ is zero for any x. All one can talk about is the probability that X takes a value in the interval between two values a and b, $P(a < X \leq b)$. The limits on this range can be infinite; for example, the probability $P(X \leq b)$ is, in effect, $P(-\infty < X \leq b)$.

The probability that a continuous random variable lies within a range can be calculated from the *probability density function* $f(x)$. This function is such that the probability that X falls between any two points a and b is equal to the area under its graph between the points a and b (Figure A.1):

$$P(a < X \leq b) = \text{Area under } f(x) \text{ between } a \text{ and } b. \qquad (A.16)$$

The density function is large in places where the random variable is likely to occur and small where it is not. These probabilities are often written using the integral notation from calculus, which here can be treated as synonymous with the area statement (see footnote 3 on page 13):

$$P(a < X \leq b) = \int_a^b f(x)\,dx. \qquad (A.17)$$

Another way to represent both discrete and continuous random variables (and some mixed types) is by their *cumulative distribution function*

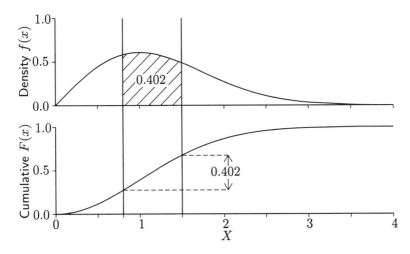

Figure A.1: *The density function and the cumulative distribution function of a continuous random variable. The probability $P(0.8 < X \leq 1.5) = 0.402$ is equal to the area under the density function between these points or the difference in the heights of the cumulative distribution function.*

$F(x)$, which gives the probability of an observation at or below its argument: $F(x) = P(X \leq x)$. For discrete and continuous random variables, respectively,

$$F(x) = \sum_{y \leq x} P(X = y) \qquad \text{and} \qquad F(x) = \int_{-\infty}^{x} f(y)\, dy. \qquad (A.18)$$

The probability that the variable falls in an interval is the difference in the cumulative distribution function at the endpoints of the interval (Figure A.1):

$$P(a < X \leq b) = F(b) - F(a). \qquad (A.19)$$

Multivariate distributions

Where several random variables are considered together, they can either be treated separately or as a *multivariate random variable*. In this book, multivariate variables are indicated by boldface letters. So the individual random variables X and Y can be combined into the bivariate random variable $\mathbf{X} = (X, Y)$. When the value taken by one variable depends on the value taken by the other, the distribution of $\mathbf{X}$ cannot be determined from the separate distributions of X and Y. A full specification requires their *joint distribution.* For discrete and continuous random variables, respectively, the joint distribution is described by a joint probability function or a joint density function:

$$P(\mathbf{X}=\mathbf{x}) = P[(X=a) \text{ and } (Y=b)] \quad \text{or} \quad f(\mathbf{x}) = f(x,y). \quad \text{(A.20)}$$

In general, the full form of these bivariate functions must be given.

A multivariate distribution is more difficult to illustrate than a univariate distribution. An n-dimensional density function needs $n+1$ dimensions to be diagrammed, n for $\mathbf{x}$ and one more for the density. A bivariate distribution forms a three-dimensional hill such as those shown in Figure 10.1 on page 173. It is usually easiest to represent such distributions by drawing the *isodensity contours* or *level curves* created by slicing through the density hill at fixed heights, as on a topographic map. For many distributions, one contour suffices to characterize the distribution. It can be chosen either to cut off a fixed proportion of the volume, say 50%, or so as to cut the axes a fixed number of standard deviations away from the center. In this book contours at one standard deviation are used.

A multivariate distribution can be reduced to a univariate distribution in several ways. One possibility is to simply ignore the other variables. So a bivariate distribution $\mathbf{X} = (X, Y)$ is reduced to its *marginal distributions* X and Y. Another possibility is to hold all but one of the variables constant, creating a *conditional distribution* (cf. Equation A.8):

$$P[(X=a)|(Y=b)] = \frac{P[(X=a) \cap (Y=b)]}{P(Y=b)} \quad \text{or} \quad f_{x|y}(x|y) = \frac{f_{xy}(x,y)}{f_y y}.$$
$$\text{(A.21)}$$

Neither the joint nor the conditional distributions contain all the information embodied in the multivariate distributions.

The concept of independence also applies to random variables. Two random variables are independent when the value taken by one variable does not influence the value taken by the other variable. The joint distribution then is the product of the marginal distributions (cf. Equation A.14). For independent discrete distributions the probabilities multiply:

$$P[(X=a) \cap (Y=b)] = P(X=a) \times P(Y=b). \quad \text{(A.22)}$$

For independent continuous random variables, the densities multiply; using subscripts to distinguish which density function is which,

$$f_{xy}(x,y) = f_x(x) f_y(y). \quad \text{(A.23)}$$

A random variable with independent components has the convenient property that the marginal and conditional distributions are the same:

$$P(X=a|Y=b) = P(X=a) \quad \text{or} \quad f_{x|y}(x|y) = f_x(x). \quad \text{(A.24)}$$

Moments and expected values

The distribution of a random variable is characterized by a series of quantities known as its *moments*, which summarize its distribution's properties. The most important of these moments is the *mean*, usually indicated by the Greek letter μ (mu), which locates the center of the distribution. The mean is calculated by summing the values that the random variable can take, weighted by their probabilities. For a discrete distribution, the mean is

$$\mu = \sum_x xP(X{=}x). \tag{A.25}$$

The sum includes all valid values of x. For a continuous distribution, this sum is replaced by an integral (which in this application cannot easily be thought of as an area) and the probability function by the density function:

$$\mu = \int_x xf(x)\,dx. \tag{A.26}$$

This integral extends over the full range of X.

Somewhat more abstractly, the mean is also known as the *expected value* or *expectation* of the random variable. As such, it is denoted by $E(X)$. Expectations can be found for other functions of X simply by replacing the multiplier x in Equations A.25 and A.26 by whatever other function is appropriate. For the function $g(x)$, the expectation is

$$E[g(x)] = \sum_x g(x)P(X{=}x) \quad \text{or} \quad E[g(x)] = \int g(x)f(x)\,dx. \tag{A.27}$$

The other moments of X are expected values of this type. The most important of these moments is the *variance*, which is the expected squared deviation from the mean. It is denoted either by σ^2 (sigma) or by $\mathrm{var}(X)$. Putting $g(x) = (x - \mu)^2$ in Equations A.27 gives

$$\sigma^2 = E[(X - \mu)^2] = \sum_x (x - \mu)^2 P(X{=}x) \quad \text{or} \quad \int (x - \mu)^2 f(x)\,dx. \tag{A.28}$$

The variance is a measure of the extent to which the distribution is spread out around the mean. It is often more convenient to represent the spread of a distribution by the *standard deviation* σ, which is the square root of the variance. The standard deviation has the advantage of being expressed in the same units as x (not their square), so it can be marked off on a histogram of the probability distribution or a plot of the density function.

Some algebra (using Equation A.32, below) shows that the variance can also be written, in what is often a better form for computation, as the difference between the expectation of X^2 and the square of the mean:

$$\sigma^2 = \mathsf{E}(X^2) - [\mathsf{E}(X)]^2. \tag{A.29}$$

Other moments besides the mean and the variance include the skew and the kurtosis. These are defined in statistics or probability texts. They play little role in the models discussed here.

In a bivariate or multivariate random variable, one further moment is important. The values taken by the two components of the bivariate random variable $\mathbf{X} = (X, Y)$ may be unrelated to each other or may tend to vary together. A measure of their common variation is the *covariance*, which is the expected value of the product of their deviations from their means:

$$\mathrm{cov}(X, Y) = \mathsf{E}[(X - \mu_X)(Y - \mu_Y)]. \tag{A.30}$$

The covariance is positive when big values of one random variable tend to go with big values of the other, and is negative when big values of one tend to go with small values of the other. It is zero when the variables are independent of each other, although zero covariance does not imply independence, except in special cases such as the Gaussian distribution discussed below. The covariance is large in magnitude (either positive or negative) when the variables are strongly related, but also when they have large variances. To standardize the measure and remove the dependence on the variability, it can be divided by the two standard deviations to give the *correlation*, denoted by ρ_{xy} (the Greek letter rho):

$$\mathrm{corr}(X, Y) = \rho_{\mathsf{xy}} = \frac{\mathrm{cov}(X, Y)}{\sigma_X \sigma_Y}. \tag{A.31}$$

The correlation equals 1 when X and Y are perfectly related in a high-with-high, low-with-low direction, and it equals -1 for a perfect reversed relationship. The correlation may be more familiar in its sample counterpart, the *Pearson correlation coefficient*, usually denoted by r.

Combinations of random variables

Many probabilistic models involve the combination of entities represented by random variables, such as sums or differences, and it is necessary to determine the distribution of these combinations. Finding these distributions in general is difficult, but fortunately only a few relationships with known distributions are important here. Particularly useful is the relationship between the moments of the original and the derived distributions.

A *linear combination* of random variables is the sum of its components, each multiplied by a constant. The linear combination of two random variables X and Y with coefficients a and b is

$$W = aX + bY.$$

Special cases include the sum (when $a = b = 1$) and the difference (when $a = 1$ and $b = -1$). The expected value of the combination is simply the combination of the expected values:

$$\mathsf{E}(W) = a\mathsf{E}(X) + b\mathsf{E}(Y) \qquad \text{or} \qquad \mu_W = a\mu_X + b\mu_Y. \qquad (\text{A}.32)$$

Consequently, the mean of the sum of random variables is the sum of their means, and the mean of their difference is the difference of their means.

The situation with variances is a little more complex. When the random variables are independent of each other, the variance of the combination is the sum of the variances, weighted by the squares of their coefficients:

$$\operatorname{var}(W) = \operatorname{var}(aX + bY) = a^2 \operatorname{var}(X) + b^2 \operatorname{var}(Y) = a^2\sigma_X^2 + b^2\sigma_Y^2. \ (\text{A}.33)$$

As this equation implies, the variance of either the sum or the difference of two independent random variables is the sum of their variances. When the variables are not independent, the covariance of the variables becomes involved:

$$\begin{aligned} \operatorname{var}(W) &= a^2 \operatorname{var}(X) + b^2 \operatorname{var}(Y) + ab \operatorname{cov}(X, Y) \\ &= a^2\sigma_X^2 + b^2\sigma_Y^2 + ab\sigma_X\sigma_Y\rho_{xy}. \end{aligned} \qquad (\text{A}.34)$$

When variables are correlated, the variance of their sum is not the same as that of their difference.

A.3 Some specific distributions

There are many standard forms of random variables, with properties that have been well worked out. Several of these figure in this book.

A shorthand for specifying that a random variable has one of the standard distributions is handy. The symbol $\sim$ stands for "is distributed as," and is followed by some symbol representing a standard form. For example, the Gaussian distribution, discussed below, is symbolized by an expression such as $\mathcal{N}(0, 1)$ or $\mathcal{N}(\mu, \sigma^2)$. The expression $X \sim \mathcal{N}(a, b)$ indicates that the random variable X is distributed according to a Gaussian distribution with parameters a and b.

The binomial and multinomial distributions

These discrete distributions arise when a fixed set of entities is distributed among a fixed set of categories. Their foundation is the *Bernoulli event*, which is simply an uncertain random event that, given an opportunity, may or may not occur. For convenience call one of these outcomes a "success" and the other a "failure." Let π be the probability of a success, and thus $1 - \pi$ is the probability of a failure. The Bernoulli event provides a description of an observation that must fall into one of two categories.

When N independent and identically distributed Bernoulli events are observed, the number of successes (or failures) has a *binomial distribution*. Under this distribution, the probability of x successes is

$$P(x) = \frac{N!}{x!(N - x)!}\pi^x(1 - \pi)^{N-x}. \tag{A.35}$$

The mean and variance of the binomial distribution are

$$\mu = N\pi \quad \text{and} \quad \sigma^2 = N\pi(1 - \pi). \tag{A.36}$$

When the entities are classified into more than two categories, the resulting random variable is multivariate; for a categories, $\mathbf{X} = (X_1, X_2, \ldots, X_a)$, with X_j counts in category j. The marginal probability of each X_j is still binomial, but, taken together, the counts have a *multinomial distribution*. When the probability that an entity is classified in category j is π_j, the probability of a particular distribution of frequencies is

$$P(x_1, x_2, \ldots, x_a) = \frac{N!}{x_1!x_2!\ldots x_a!}\pi_1^{x_1}\pi_1^{x_2}\ldots\pi_a^{x_a}. \tag{A.37}$$

Because each of the X_j has a binomial distribution, the mean and the variance of each of the counts in a multinomial distribution are

$$\mathsf{E}(X_j) = N\pi_j \quad \text{and} \quad \text{var}(X_j) = N\pi_j(1 - \pi_j). \tag{A.38}$$

However, the counts are not independent. The total number of entities is fixed at N, so a large number of observations in one category must be accompanied by smaller numbers in the other categories. The covariance of any two counts is

$$\text{cov}(X_j, X_k) = -N\pi_j\pi_k. \tag{A.39}$$

In many applications, one works with proportions instead of frequencies. The proportion $P_j = X_j/N$ is a random variable and has a distribution that is multinomial, like that of X_j, except that its mass falls at the points 0,

$1/N$, $2/N$, and so forth, instead of at the integers. The mean of P_j is equal to the population probability, $E(P_j) = \pi_j$, and its variability is

$$\text{var}(P_j) = \frac{\pi_j(1 - \pi_j)}{N}, \qquad \text{and} \qquad \text{cov}(P_j, P_k) = \frac{-\pi_j\pi_k}{N}. \qquad \text{(A.40)}$$

When the probabilities π_j are estimated by the observed proportions $p_j = x_j/N$, those values are used to estimate the variability:

$$\widehat{\text{var}}(P_j) = \frac{p_j(1 - p_j)}{N} \qquad \text{and} \qquad \widehat{\text{cov}}(P_j, P_k) = \frac{-p_jp_k}{N}. \qquad \text{(A.41)}$$

The Gaussian (or normal) distribution

Certainly the most important distribution for signal-detection theory is the *Gaussian distribution*, also known as the *normal distribution*. The name commemorates the great mathematician Karl Friedrich Gauss (1777–1855), who developed many of its properties as an error distribution. It is more often known as the normal distribution in statistical work, but the term Gaussian is more usual in signal-detection theory. Because of its importance, this section describes the Gaussian distribution in some detail.

The Gaussian distribution is defined over all real numbers, and it has the bell-shaped density function ubiquitous in statistics. There is actually a family of Gaussian distributions, each member of which depends on two parameters, the center (or mean) and the spread (or variance). Gaussian distributions are denoted by the expression $\mathcal{N}(\mu, \sigma^2)$. Note that the variance σ^2 is used here, not the standard deviation.

It is useful to get a feeling for the numerical placement of the distribution on its axis. Figure A.2 shows the density functions of three Gaussian distributions with different means and variances. About $2/3$ (actually 68.2%) of each of these distributions falls within one standard deviation of the mean, about 95% within two standard deviations, and essentially all of it within three standard deviations.

The density function of the general $\mathcal{N}(\mu, \sigma^2)$ Gaussian distribution is[1]

$$f(x) = \frac{1}{\sqrt{2\pi\sigma^2}} \exp\left[-\frac{1}{2}\frac{(x - \mu)^2}{\sigma^2}\right]. \qquad \text{(A.42)}$$

Most calculations with the Gaussian distribution refer to areas under this function.

---

[1] The function $\exp(a)$ is a synonym for the exponential e^a, where $e = 2.718281828\ldots$ is a fundamental mathematical constant and the base of natural logarithms. The exp notation is more convenient than a superscript when the exponent is complicated. The two forms are used interchangeably below—for example, compare Equations A.42 and A.43.

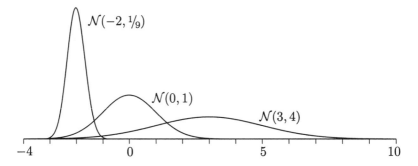

Figure A.2: The density functions for several members of the family of Gaussian distributions with different means and variances.

From a computational point of view, the key member of the Gaussian family is the $\mathcal{N}(0,1)$ distribution, with mean of zero and variance of one. The density function of this *standard Gaussian distribution* is denoted by the lowercase Greek letter phi:[2]

$$\varphi(z) = \frac{1}{\sqrt{2\pi}} e^{-z^2/2}. \tag{A.43}$$

Its cumulative distribution function is indicated by the corresponding uppercase letter:

$$\Phi(z) = \int_{-\infty}^{z} \varphi(x)\, dx. \tag{A.44}$$

One often has need of the inverse of the cumulative Gaussian function—a transformation that takes a probability p and converts it to a standard score z. This function is denoted here by $Z(p)$—it could be written in standard mathematical notation as the inverse function $\Phi^{-1}(p)$ (see footnote 1 on page 45), but it is more convenient (and less intimidating) to give this inverse its own symbol. Thus, $z = Z(p)$ means that $p = \Phi(z)$. Numerically, $\Phi(0.3) = 0.618$ and $Z(0.618) = 0.3$.

The Gaussian distribution is symmetrical about its mean, so that the density function of the standard distribution is the same for positive and negative arguments:

$$\varphi(z) = \varphi(-z).$$

Because of this symmetry, the area in the lower tail of the density function below the value z is equal to the area in the upper tail above the value $-z$:

---

[2] In this book an old-style phi, φ, is used for the Gaussian density. You may find the newer form ϕ used in some references.

An equivalent form of this relationship is

$$\Phi(z) = 1 - \Phi(-z). \tag{A.45}$$

Neither the cumulative Gaussian distribution $\Phi(z)$ nor its inverse $Z(p)$ has any finite expression in terms of "elementary" operations such as addition, multiplication, exponentiation, logarithms, and the like. However, it is quite possible to calculate these areas numerically. There are several computer approximations to both functions, and they are now available on several scientific calculators. The easiest way to get values for hand calculation is to use a table, such as that found in practically every statistics book. Two such tables are provided below, starting on page 249 at the end of this appendix. These tables give the standard probabilities $\Phi(z)$ and their inverse $Z(p)$ for a range of values of z and p. The rows of the tables are indexed by the first two digits of z or p, and the columns by the third digit. Both tables are accurate to the three places given.

Example A.1: Find the probability of a score below the standard score of -1.24.

Solution: First sketch a figure and shade the area that is wanted—always a good idea to avoid errors in the sign of z or of confusing errors in upper or lower tails:

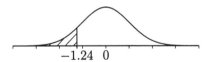

The figure shows that the answer should be less than 0.5. The problem asks for $\Phi(-1.24)$, but the tables have entries only for $z > 0$. However, by Equation A.45, $\Phi(-1.24) = 1 - \Phi(1.24)$, and the latter value is in the tables. Enter the first table in the row labeled 1.20. Select the fifth column (headed 0.04) and read off the probability 0.893. Then $\Phi(-1.24) = 1 - 0.893 = 0.107$.

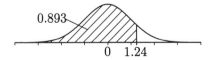

Example A.2: What value in a standard Gaussian distribution is exceeded 28.5% of the time?

Solution: Again begin with a figure, with a division at z cutting off about 28.5% of the distribution:

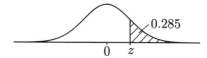

The figure indicates that z will be positive. With 28.5% above z, there is 71.5% below. Enter the table of $Z(p)$ on page 250 in the row labeled 0.710 and the column labeled 0.005 to get the answer $z = 0.568$.

The importance of the standard distribution lies in the fact that every member of the normal family has a the same general shape and so can be converted to the standard distribution by a linear transformation. Hence, questions relating to probabilities for a $\mathcal{N}(\mu, \sigma^2)$ distribution or positions within that distribution can be answered by restating the questions in terms of the standard $\mathcal{N}(0, 1)$ distribution. Both the density function $f(x)$ and the cumulative distribution function $F(x)$ of a general Gaussian distribution are converted to standard form by subtracting the mean from the observation x and dividing the result by the standard deviation. The probability of an observation below x in a $\mathcal{N}(\mu, \sigma^2)$ distribution is

$$F(x) = \Phi(z) \quad \text{with} \quad z = \frac{x - \mu}{\sigma}, \tag{A.46}$$

and the position in that distribution with an area p below it is

$$x = \sigma Z(p) + \mu. \tag{A.47}$$

One must be careful using these transformations to get heights of density functions, as a factor of $1/\sigma$ appears in the general density that disappears when $\sigma = 1$ (compare Equations A.42 and A.43). When in doubt, calculate densities directly from Equation A.42.

Example A.3: With $X \sim \mathcal{N}(4, 6)$, what is $P(X < 7)$? What value of X has 60% of the distribution below it?

Solution: Convert the nonstandard distribution to a standard one. The desired area is the same as $\Phi(z)$ with

$$z = \frac{x - \mu}{\sigma} = \frac{7 - 4}{\sqrt{6}} = 1.225.$$

From the table of $\Phi(z)$ on page 249, $\Phi(1.22) = 0.889$ and $\Phi(1.23) = 0.891$. Thus, $P(X < 7) = 0.890$.

From the table of $Z(p)$ on page 250, the score $z = 0.253$ has 60% of the distribution below it. The corresponding unstandardized value is

$$x = \mu + \sigma z = 4 + \sqrt{6}z = 4 + \sqrt{6} \times 0.253 = 4.62.$$

When several Gaussian random variables are combined in a linear combination, the result also has a Gaussian distribution. The parameters of the new distribution are given by Equations A.32–A.34. This very convenient property—linear combinations stay within the Gaussian family—is not shared by the other distributions discussed here.

Gaussian distributions are widely used, both in statistics and as a model for uncertain measurements. The most important reason for its importance in signal-detection theory is its appearance as the limit of the sum of independent observations. The *central limit theorem* states that the distribution of the sum of a series of similar random variables tends to the Gaussian form when the number of terms gets large. In more detail, suppose that $X_1, X_2, \ldots, X_N$ are mutually independent random variables with the same distribution (there are a few other conditions that are not important here). Calculate their sum Y. When $N = 1$, Y has the same distribution as X. As terms are added to the sum, any irregularities in the distribution become smoothed out. As terms are added, both the mean and the variance of Y increase—by Equations A.32 and A.33, $\mu_Y = N\mu_X$ and $\sigma_Y^2 = N\sigma_X^2$. However, the central limit theorem states that if Y is standardized by subtracting its mean and dividing by its standard deviation (as in Equation A.46), then as the number of observations increases, the result approaches a standard $\mathcal{N}(0, 1)$ Gaussian distribution. In practice, the number of terms needed to give a smooth and symmetrical distribution is not usually very large. When even a moderate number of random quantities are combined—and other versions of the theorem show they need not be either exactly identical or quite independent—their sum is often remarkably well approximated by a Gaussian distribution.[3]

The logistic distribution

The *logistic distribution* has a symmetrical, bell-shaped density function very similar to that of a Gaussian distribution. It has sometimes been used instead of the Gaussian in signal-detection theory to describe the underlying random variables on which the decision is based. Again like the Gaussian distribution, there is a family of logistic distributions, differing in center and spread, but having the same general shape. The family depends on

---

[3] Another reason for the ubiquity of the Gaussian distribution is its relationship to the sampling distribution of the sums of squares. This result, which is fundamental to the analysis of variance, is not so important to signal-detection theory.

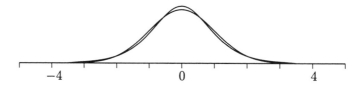

Figure A.3: Density functions for a standard Gaussian distribution and for the logistic distribution that matches it most closely ($\mu = 0$ and $\tau = {}^{10}\!/_{17}$). The logistic distribution has the greater height at the center.

two parameters, a mean μ and a scale parameter τ. The logistic density function is

$$f(x) = \frac{e^{-(x-\mu)/\tau}}{\tau[1 + e^{-(x-\mu)/\tau}]^2}. \tag{A.48}$$

Although the parameter τ governs the spread of the distribution, it is not equal to the standard deviation. The variance of the logistic distribution is

$$\text{var}(X) = \tfrac{1}{3}\pi^2\tau^2, \qquad \text{so} \qquad \sigma = \pi\tau/\sqrt{3}. \tag{A.49}$$

An important difference between the logistic distribution and the Gaussian is that the logistic cumulative distribution function has a comparatively simple form:

$$F(x) = \frac{1}{1 + e^{-(x-\mu)/\tau}}. \tag{A.50}$$

In standard form, with $\mu = 0$ and $\tau = 1$, the cumulative function can be inverted to give

$$x = \text{logit}[F(x)]. \tag{A.51}$$

Equations A.50 and A.51 give the logistic distribution a computational tractability that the Gaussian distribution lacks.

The logistic and Gaussian distributions have very similar shapes. The member of the logistic family that best matches the standard Gaussian distribution has $\mu = 0$ and $\tau = 16\sqrt{3}/15\pi = 0.5881$, a value that is closely approximated by ${}^{10}\!/_{17}$. With these values, the cumulative distribution functions of the two distributions do not differ by more than 0.01, an amount that is usually less than the precision of most sets of data. Figure A.3 shows the density function of a standard Gaussian distribution and a logistic distribution with $\mu = 0$ and $\tau = {}^{10}\!/_{17}$. The similarity of the two is obvious.

The difference between the Gaussian and logistic distributions is small enough that in most applications either can be used. The Gaussian distribution has a stronger theoretical basis in the central limit theorem, while the closed form of its cumulative distribution function makes the logistic distribution somewhat easier to use analytically. The computational advantages of the logistic distribution are less important now than they once were, so, because of its theoretical strengths and its greater use historically, the Gaussian form is emphasized in this book.

The multivariate Gaussian distribution

The univariate Gaussian distribution extends naturally to multivariate random variables. Most of the characteristics of the multivariate distribution are present in its bivariate form, which is emphasized here. A random variable $\mathbf{X} = (X_1, X_2)$ has a *bivariate Gaussian distribution* when each of its components is Gaussian and the association between them has the particularly simple form described below.

There is a family of bivariate Gaussian distributions that vary in their centers and their variability. Each member of the family is centered at a bivariate mean $\boldsymbol{\mu} = (\mu_1, \mu_2)$. The variability is determined by the variances σ_1^2 and σ_2^2 of the individual components (or, equivalently, their standard deviations) and by the correlation ρ between them. For compactness, the variance parameters are usually combined in a *covariance matrix*, denoted by an uppercase Greek letter sigma (used here as a symbol, not a summation sign):

$$\boldsymbol{\Sigma} = \begin{bmatrix} \sigma_1^2 & \sigma_1\sigma_2\rho \\ \sigma_1\sigma_2\rho & \sigma_2^2 \end{bmatrix}.$$

In a generalization of the univariate notation, one indicates that $\mathbf{X}$ has a Gaussian distribution with mean $\boldsymbol{\mu}$ and covariance matrix $\boldsymbol{\Sigma}$ by writing $\mathbf{X} \sim \mathcal{N}(\boldsymbol{\mu}, \boldsymbol{\Sigma})$. This representation carries over to higher-dimensional forms of the distribution. For example, a quadravariate (four variable) Gaussian distribution has four means, four variances, one for each variable, and six correlations, one for each pair of variables.

The bivariate Gaussian density has the bell-shaped form shown on the left in Figure 10.1 on page 173. The level curves of a Gaussian distribution are ellipses (including circles). Only such contours are possible for Gaussian distributions. Distributions with contours shaped like bananas, splotches, or other forms, regular or irregular, are not Gaussian, even when the univariate distributions of their components are Gaussian.

When the two components of a bivariate Gaussian distribution have equal variance and are uncorrelated, the contour ellipses are circles. When the variances are unequal and the components are uncorrelated, the contours are ellipses with their long dimension parallel to the axis of the most variable component. The widths of the ellipse, when cut parallel to one of the axes, is proportional to the corresponding standard deviations. When the components are correlated, the long axis of the ellipse lies at an angle to the component axes, from lower left to upper right when the correlation is positive and the reverse when it is negative.

Like the univariate density function (Equation A.42), the multivariate Gaussian density function is the product of two terms. One is a constant C_{Σ} that depends only on the covariance matrix; the other is an exponential that contains a quadratic function $Q_{\Sigma}(\mathbf{x})$ of the deviations of the observation from the mean:

$$f(\mathbf{x}) = C_{\Sigma} \exp\left[-\tfrac{1}{2}Q_{\Sigma}(\mathbf{x} - \boldsymbol{\mu})\right].$$

In two dimensions, the two functions can be written out algebraically:

$$C_{\Sigma} = \frac{1}{2\pi\sigma_1\sigma_2\sqrt{1-\rho^2}} \quad \text{and} \quad Q_{\Sigma}(\mathbf{x}) = \sigma_1^2 x_1^2 + 2\rho\sigma_1\sigma_2 x_1 x_2 + \sigma_2^2 x_2^2.$$

In higher dimensions, it is only practical to express them with matrices.[4]

The distributions with independent components play a special role in signal-detection theory. Their joint density function is the product of the individual densities (Equation A.23). With Gaussian distributions, independence gives $Q_{\Sigma}(\mathbf{x})$ a simple form, regardless of the number of dimensions. The joint distribution of d independent standard Gaussian variables is[5]

$$f(\mathbf{z}) = f_1(z_1)f_2(z_2)\ldots f_d(z_d)$$

$$= \prod_{i=1}^{d} \frac{1}{\sqrt{2\pi}} e^{-z_i^2/2}$$

$$= \frac{1}{(2\pi)^{d/2}} \exp\left[-\tfrac{1}{2}\sum_{i=1}^{d} z_i^2\right]. \qquad \text{(A.52)}$$

Fundamentally, $Q_{\Sigma}(\mathbf{x})$ here is the sum of squares.

---

[4]The matrix expressions in d dimensions, which are not needed in this book, are

$$C_{\Sigma} = [(2\pi)^{d/2}|\Sigma|]^{-1} \quad \text{and} \quad Q_{\Sigma}(\mathbf{x}) = \mathbf{x}'\Sigma^{-1}\mathbf{x}.$$

[5]The symbol $\prod$ denotes a product. It is similar to the summation symbol $\sum$, but the terms are multiplied together instead of being summed.

The chi-square distribution

An important distribution derives from the squares of Gaussian random variables. Suppose that $Z_1, Z_2, \ldots, Z_d$ are independent standard Gaussian $\mathcal{N}(0,1)$ random variables. Now square each of the variables and add them up:

$$X^2 = Z_1^2 + Z_2^2 + \cdots + Z_d^2. \tag{A.53}$$

The resulting random variable is said to have a *chi-square distribution*, symbolized χ_d^2. The number of terms entering into this sum, here d, is known as the *degrees of freedom* of the chi-square variable. Unlike Gaussian random variables, a chi-square random variable takes only positive values.

The number of degrees of freedom defines a family of chi-square random variables, but it is not a family in the sense that the Gaussian or logistic families contain members with any mean and variance. The mean and variance of a chi-square random variable are completely determined by its degrees of freedom (they are d and $2d$, respectively). Moreover, unlike Gaussian distributions that have the same bell-shaped form regardless of their mean and variance, the shape of the chi-square distribution depends on its degrees of freedom (Figure A.4). The distribution with one degree of freedom is highly skewed—its density function has a mode at zero. With more terms in the sum, the chi-square density becomes more rounded. When there are many degrees of freedom, the central limit theorem implies that the shape of the chi-square distribution is roughly Gaussian.

The chi-square distribution is important in signal-detection theory in two distinct domains. One application is as the distribution associated with one or both of the signal and noise events. It occurs here in decisions based on signal energy, which is the square of signal magnitude. This application is touched on only briefly in this book (see Section 9.4 and Figure 9.5). The other application is statistical. Many tests of the agreement between data and a theoretical description produce test statistics that have a chi-square distribution.

Consider an experiment in which each of a series of events—they may be trials in an experiment, or the results from individual subjects—is classified into one and only one of a set of c categories.[6] The result is a series of frequencies, $x_1, x_2, \ldots, x_c$, indicating how many events are in each category. For example, the responses of a subject may be classified into the categories *correct* or *error*. Now suppose that a theoretical model has been fitted to the data, which makes predictions of the mean number of counts

---

[6]In many applications, the data form a two-way table or other configuration, such as a contingency table. However, the calculation and interpretation of the test require only that the observations have been categorized somehow.

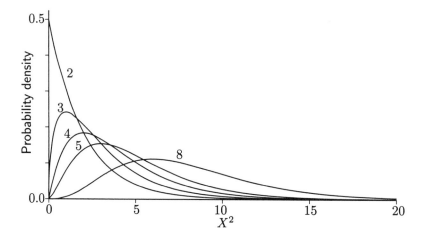

Figure A.4: *The density functions of chi-square distributions with several different degrees of freedom. A chi-square density function with one degree of freedom is shown as the distribution of Y_n in Figure 9.5 on page 168. Distributions with larger degrees of freedom approach the Gaussian in form.*

in each category, $m_1, m_2, \ldots, m_c$. These means need not be integers. One would like to decide whether the model fits the data well or poorly, that is, whether the x_i deviate from the m_i by an improbably large amount, were the model correct. The test is made by calculating one of two measures of the discrepancy between the data and the predictions, the *Pearson statistic,*

$$X^2 = \sum_{i=1}^{c} \frac{(x_i - m_i)^2}{m_i}, \tag{A.54}$$

and the *likelihood-ratio statistic,*

$$G^2 = 2 \sum_{i=1}^{c} x_i \log \frac{x_i}{m_i}. \tag{A.55}$$

Both statistics measure badness of fit, in the sense that big values arise when the model fits the data poorly (how that happens is clear from the squared difference in the numerator of the Pearson statistic, but is opaque for the likelihood-ratio statistic). Both X^2 and G^2 are small when the model and the data agree.

To get an indication of whether a model is satisfactory, one can calculate either of these badness-of-fit statistics and see whether it is large enough to have been unlikely to have arisen by chance. Here is where the chi-square distribution comes in. Suppose that the counts can be deemed to satisfy two assumptions (for details, see a text on general statistics or categorical data analysis). First, the observations are independent of each other, and, second, they are probabilistically identical. Now, if the model is correct and the sample size is sufficiently great to make the m_j reasonably large, then both X^2 and G^2 have roughly a chi-square distribution. The number of observations should be large enough that almost all the m_{ij} are larger than a half a dozen (although a few cells in larger tables can be smaller than this). Values of either statistic that are too large fall in the upper tail of the distribution and are grounds to reject the model. The degrees of freedom for this test depend on three factors. Each category adds a degree of freedom, each constraint on the data takes one away, as does each parameter estimated to fit the model:

$$df = (\text{categories}) - (\text{constraints}) - (\text{parameters}). \qquad (A.56)$$

The constraints arise from the structure of the study, usually because the number of observations is fixed. For example, a detection study with 80 signal trials and 80 noise trials imposes two constraints, one on the number of hits and misses, the other on the number of false alarms and correct rejections. When the hypothesis being tested is of independence of the rows and columns of a two-way table, the rule leads to the familiar formula

$$df = (\text{rows} - 1)(\text{columns} - 1). \qquad (A.57)$$

To complete the test, the cumulative chi-square distribution is needed. The table on page 251 gives critical values from this distribution. It shows, for distributions with up to 30 degrees of freedom, values such that the upper tail contains fixed percentages of the distribution. For example, the table shows that 10% of a chi-square distribution with 2 degrees of freedom falls above the value 4.16 and that one quarter of a chi-square distribution with 10 degrees of freedom falls above 9.34.

Example A.4: Sixty subjects are classified into four categories and fitted with a one-parameter model. The observed and fitted frequencies are

x_i	20	16	12	12

m_i	28	15	10	7

Is the model satisfactory?

Solution: The Pearson statistic and likelihood-ratio statistics (Equations A.54 and A.55) are

$$X^2 = \frac{(20 - 28)^2}{28} + \frac{(16 - 15)^2}{15} + \frac{(12 - 10)^2}{10} + \frac{(12 - 7)^2}{7} = 6.32,$$

$$G^2 = 2 \left[20 \log \frac{20}{28} + 16 \log \frac{16}{15} + 12 \log \frac{12}{10} + 12 \log \frac{12}{7} \right] = 5.92.$$

There are four cells in the table, one constraint (note that both the x_i and the m_i sum to 60, the number of observations), and the model contains one parameter. By Equation A.56, the test statistics have $4 - 1 - 1 = 2$ degrees of freedom, and from the table on page 251, 5% of a chi-square distribution with 2 degrees of freedom falls above 5.99. The observed statistics bracket this figure; if X^2 were used, the model would be rejected, and if G^2 were used, it would barely be retained. As this result points out, the interpretation sometimes depends on the statistic used, and values close to any criterion should be treated with caution.

As between the two statistics, X^2 is usually preferable to G^2 for simple tests of the type described in Example A.4, not least because of its greater familiarity. The statistic G^2 has greater value in the comparison of different models, an application that is discussed in Section 11.6.

Table of $\Phi(z)$ for the standard Gaussian distribution

	0.00	0.01	0.02	0.03	0.04	0.05	0.06	0.07	0.08	0.09
0.00	0.500	0.504	0.508	0.512	0.516	0.520	0.524	0.528	0.532	0.536
0.10	0.540	0.544	0.548	0.552	0.556	0.560	0.564	0.567	0.571	0.575
0.20	0.579	0.583	0.587	0.591	0.595	0.599	0.603	0.606	0.610	0.614
0.30	0.618	0.622	0.626	0.629	0.633	0.637	0.641	0.644	0.648	0.652
0.40	0.655	0.659	0.663	0.666	0.670	0.674	0.677	0.681	0.684	0.688
0.50	0.691	0.695	0.698	0.702	0.705	0.709	0.712	0.716	0.719	0.722
0.60	0.726	0.729	0.732	0.736	0.739	0.742	0.745	0.749	0.752	0.755
0.70	0.758	0.761	0.764	0.767	0.770	0.773	0.776	0.779	0.782	0.785
0.80	0.788	0.791	0.794	0.797	0.800	0.802	0.805	0.808	0.811	0.813
0.90	0.816	0.819	0.821	0.824	0.826	0.829	0.831	0.834	0.836	0.839
1.00	0.841	0.844	0.846	0.848	0.851	0.853	0.855	0.858	0.860	0.862
1.10	0.864	0.867	0.869	0.871	0.873	0.875	0.877	0.879	0.881	0.883
1.20	0.885	0.887	0.889	0.891	0.893	0.894	0.896	0.898	0.900	0.901
1.30	0.903	0.905	0.907	0.908	0.910	0.911	0.913	0.915	0.916	0.918
1.40	0.919	0.921	0.922	0.924	0.925	0.926	0.928	0.929	0.931	0.932
1.50	0.933	0.934	0.936	0.937	0.938	0.939	0.941	0.942	0.943	0.944
1.60	0.945	0.946	0.947	0.948	0.949	0.951	0.952	0.953	0.954	0.954
1.70	0.955	0.956	0.957	0.958	0.959	0.960	0.961	0.962	0.962	0.963
1.80	0.964	0.965	0.966	0.966	0.967	0.968	0.969	0.969	0.970	0.971
1.90	0.971	0.972	0.973	0.973	0.974	0.974	0.975	0.976	0.976	0.977
2.00	0.977	0.978	0.978	0.979	0.979	0.980	0.980	0.981	0.981	0.982
2.10	0.982	0.983	0.983	0.983	0.984	0.984	0.985	0.985	0.985	0.986
2.20	0.986	0.986	0.987	0.987	0.987	0.988	0.988	0.988	0.989	0.989
2.30	0.989	0.990	0.990	0.990	0.990	0.991	0.991	0.991	0.991	0.992
2.40	0.992	0.992	0.992	0.992	0.993	0.993	0.993	0.993	0.993	0.994
2.50	0.994	0.994	0.994	0.994	0.994	0.995	0.995	0.995	0.995	0.995
2.60	0.995	0.995	0.996	0.996	0.996	0.996	0.996	0.996	0.996	0.996
2.70	0.997	0.997	0.997	0.997	0.997	0.997	0.997	0.997	0.997	0.997
2.80	0.997	0.998	0.998	0.998	0.998	0.998	0.998	0.998	0.998	0.998
2.90	0.998	0.998	0.998	0.998	0.998	0.998	0.998	0.999	0.999	0.999
3.00	0.999	0.999	0.999	0.999	0.999	0.999	0.999	0.999	0.999	0.999
3.10	0.999	0.999	0.999	0.999	0.999	0.999	0.999	0.999	0.999	0.999
3.20	0.999	0.999	0.999	0.999	0.999	0.999	0.999	0.999	0.999	0.999
3.30	1.000	1.000	1.000	1.000	1.000	1.000	1.000	1.000	1.000	1.000
3.40	1.000	1.000	1.000	1.000	1.000	1.000	1.000	1.000	1.000	1.000
3.50	1.000	1.000	1.000	1.000	1.000	1.000	1.000	1.000	1.000	1.000

Table of $Z(p)$ for the standard Gaussian distribution

	0.000	0.001	0.002	0.003	0.004	0.005	0.006	0.007	0.008	0.009
0.500	0.000	0.003	0.005	0.008	0.010	0.013	0.015	0.018	0.020	0.023
0.510	0.025	0.028	0.030	0.033	0.035	0.038	0.040	0.043	0.045	0.048
0.520	0.050	0.053	0.055	0.058	0.060	0.063	0.065	0.068	0.070	0.073
0.530	0.075	0.078	0.080	0.083	0.085	0.088	0.090	0.093	0.095	0.098
0.540	0.100	0.103	0.105	0.108	0.110	0.113	0.116	0.118	0.121	0.123
0.550	0.126	0.128	0.131	0.133	0.136	0.138	0.141	0.143	0.146	0.148
0.560	0.151	0.153	0.156	0.159	0.161	0.164	0.166	0.169	0.171	0.174
0.570	0.176	0.179	0.181	0.184	0.186	0.189	0.192	0.194	0.197	0.199
0.580	0.202	0.204	0.207	0.209	0.212	0.215	0.217	0.220	0.223	0.225
0.590	0.228	0.230	0.233	0.235	0.238	0.240	0.243	0.246	0.248	0.251
0.600	0.253	0.256	0.259	0.261	0.264	0.266	0.269	0.271	0.274	0.277
0.610	0.279	0.282	0.284	0.287	0.290	0.292	0.295	0.298	0.300	0.303
0.620	0.305	0.308	0.311	0.313	0.316	0.319	0.321	0.324	0.327	0.329
0.630	0.332	0.335	0.337	0.340	0.342	0.345	0.348	0.350	0.353	0.356
0.640	0.358	0.361	0.364	0.366	0.369	0.372	0.374	0.377	0.380	0.383
0.650	0.385	0.388	0.391	0.393	0.396	0.399	0.402	0.404	0.407	0.410
0.660	0.412	0.415	0.418	0.421	0.423	0.426	0.429	0.432	0.434	0.437
0.670	0.440	0.443	0.445	0.448	0.451	0.454	0.456	0.459	0.462	0.465
0.680	0.468	0.470	0.473	0.476	0.479	0.482	0.484	0.487	0.490	0.493
0.690	0.496	0.499	0.501	0.504	0.507	0.510	0.513	0.516	0.519	0.521
0.700	0.524	0.527	0.530	0.533	0.536	0.539	0.542	0.544	0.547	0.550
0.710	0.553	0.556	0.559	0.562	0.565	0.568	0.571	0.574	0.577	0.580
0.720	0.583	0.586	0.589	0.592	0.595	0.598	0.601	0.604	0.607	0.610
0.730	0.613	0.616	0.619	0.622	0.625	0.628	0.631	0.634	0.637	0.640
0.740	0.643	0.646	0.649	0.652	0.656	0.659	0.662	0.665	0.668	0.671
0.750	0.674	0.678	0.681	0.684	0.687	0.690	0.693	0.697	0.700	0.703
0.760	0.706	0.709	0.713	0.716	0.719	0.722	0.726	0.729	0.732	0.735
0.770	0.739	0.742	0.745	0.749	0.752	0.755	0.758	0.762	0.765	0.768
0.780	0.772	0.775	0.779	0.782	0.785	0.789	0.792	0.796	0.799	0.803
0.790	0.806	0.810	0.813	0.817	0.820	0.824	0.827	0.831	0.834	0.838
0.800	0.842	0.845	0.849	0.852	0.856	0.859	0.863	0.867	0.870	0.874
0.810	0.878	0.881	0.885	0.889	0.893	0.896	0.900	0.904	0.908	0.911
0.820	0.915	0.919	0.923	0.927	0.931	0.934	0.938	0.942	0.946	0.950
0.830	0.954	0.958	0.962	0.966	0.970	0.974	0.978	0.982	0.986	0.990
0.840	0.994	0.999	1.003	1.007	1.011	1.015	1.020	1.024	1.028	1.032
0.850	1.036	1.041	1.045	1.050	1.054	1.058	1.063	1.067	1.072	1.076
0.860	1.080	1.085	1.090	1.094	1.099	1.103	1.108	1.113	1.117	1.122
0.870	1.126	1.132	1.136	1.141	1.146	1.151	1.156	1.161	1.166	1.171
0.880	1.175	1.181	1.186	1.191	1.196	1.201	1.206	1.211	1.217	1.222
0.890	1.227	1.233	1.238	1.243	1.249	1.254	1.258	1.264	1.270	1.275
0.900	1.282	1.287	1.292	1.298	1.304	1.310	1.316	1.322	1.328	1.334
0.910	1.341	1.346	1.353	1.359	1.365	1.372	1.378	1.385	1.391	1.398
0.920	1.405	1.412	1.418	1.425	1.432	1.439	1.446	1.454	1.461	1.468
0.930	1.476	1.483	1.491	1.498	1.507	1.515	1.523	1.531	1.539	1.547
0.940	1.555	1.564	1.573	1.581	1.590	1.599	1.608	1.617	1.627	1.636
0.950	1.645	1.656	1.666	1.676	1.686	1.697	1.707	1.718	1.729	1.741
0.960	1.751	1.764	1.776	1.785	1.798	1.811	1.824	1.837	1.851	1.865
0.970	1.881	1.895	1.910	1.926	1.942	1.959	1.976	1.994	2.013	2.033
0.980	2.054	2.074	2.096	2.119	2.144	2.170	2.197	2.226	2.257	2.290
0.990	2.326	2.365	2.409	2.457	2.512	2.576	2.652	2.748	2.879	3.093

Critical values of the chi-square distribution

ν	Upper-tail probability α					
	0.5	0.25	0.1	0.05	0.01	0.001
1	0.45	1.32	2.71	3.84	6.63	10.83
2	1.39	2.77	4.61	5.99	9.21	13.82
3	2.37	4.11	6.25	7.81	11.34	16.27
4	3.36	5.39	7.78	9.49	13.28	18.47
5	4.35	6.63	9.24	11.07	15.09	20.51
6	5.35	7.84	10.64	12.59	16.81	22.46
7	6.35	9.04	12.02	14.07	18.48	24.32
8	7.34	10.22	13.36	15.51	20.09	26.12
9	8.34	11.39	14.68	16.92	21.67	27.88
10	9.34	12.55	15.99	18.31	23.21	29.59
11	10.34	13.70	17.28	19.68	24.72	31.26
12	11.34	14.85	18.55	21.03	26.22	32.91
13	12.34	15.98	19.81	22.36	27.69	34.53
14	13.34	17.12	21.06	23.68	29.14	36.12
15	14.34	18.25	22.31	25.00	30.58	37.70
16	15.34	19.37	23.54	26.30	32.00	39.25
17	16.34	20.49	24.77	27.59	33.41	40.79
18	17.34	21.60	25.99	28.87	34.81	42.31
19	18.34	22.72	27.20	30.14	36.19	43.82
20	19.34	23.83	28.41	31.41	37.57	45.32
21	20.34	24.93	29.62	32.67	38.93	46.80
22	21.34	26.04	30.81	33.92	40.29	48.27
23	22.34	27.14	32.01	35.17	41.64	49.73
24	23.34	28.24	33.20	36.42	42.98	51.18
25	24.34	29.34	34.38	37.65	44.31	52.62
26	25.34	30.43	35.56	38.89	45.64	54.05
27	26.34	31.53	36.74	40.11	46.96	55.48
28	27.34	32.62	37.92	41.34	48.28	56.89
29	28.34	33.71	39.09	42.56	49.59	58.30
30	29.34	34.80	40.26	43.77	50.89	59.70

References

Agresti, A. (1990). *Categorical data analysis*. New York: Wiley.

Ashby, F. G. (1992). *Multidimensional models for perception and cognition*. Hillsdale, NJ: Erlbaum.

Ashby, F. G., & Townsend, J. T. (1986). Varieties of perceptual independence. *Psychological Bulletin*, *93*, 154–179.

Atkinson, R. C., Bower, G. H., & Crothers, E. J. (1965). *An introduction of mathematical learning theory*. New York: Wiley.

Dorfman, D. D., & Alf, E. A., Jr. (1968a). Maximum likelihood estimation of signal detection theory and determination of confidence intervals—rating-method data. *Journal of Mathematical Psychology*, *6*, 487–496.

Dorfman, D. D., & Alf, E. A., Jr. (1968b). Maximum likelihood estimation of signal detection theory—a direct solution. *Psychometrika*, *33*, 117–124.

Egan, J. P. (1975). *Signal detection theory and ROC analysis*. New York: Academic Press.

Falmagne, J.-C. (1985). *Elements of psychophysical theory*. New York: Oxford University Press.

Green, D. M., & Swets, J. W. (1966). *Signal detection theory and psychophysics*. New York: Wiley. (Reprint: Huntington, NY: Krieger, 1974.)

Johnson, N. L., Kotz, S., & Balakrishnan, N. (1994). *Continuous univariate distributions, Vol. 1* (second ed.). New York: Wiley.

Johnson, N. L., Kotz, S., & Balakrishnan, N. (1995). *Continuous univariate distributions, Vol. 2* (second ed.). New York: Wiley.

Johnson, N. L., Kotz, S., & Kemp, A. W. (1992). *Univariate discrete distributions* (second ed.). New York: Wiley.

Kruskal, J. B., & Wish, M. (1978). *Multidimensional scaling*. Newbury Park, CA: Sage.

Link, S. W. (1992). *The wave theory of difference and similarity*. Hillsdale, NJ: Erlbaum.

Luce, D. R. (1959). *Individual choice behavior*. New York: Wiley.

Macmillan, N. A., & Creelman, C. D. (1991). *Detection theory: a user's guide*. Cambridge: Cambridge University Press.

Macmillan, N. A., & Creelman, C. D. (1996). Triangles in ROC space: history and theory of "nonparametric" measures of sensitivity and response bias. *Psychonomic Bulletin and Review, 3*, 164–170.

McNicol, D. (1972). *A primer of signal detection theory*. London: Allen & Unwin.

Metz, C. E., & Pan, X. (1999). "Proper" binormal ROC curves: Theory and maximum-likelihood estimation. *Journal of Mathematical Psychology, 43*, 1–33.

Metz, C. E., Want, P.-L., & Kronman, H. B. (1984). A new approach for testing the significance of differences between ROC curves measured from correlated data. In F. Diconinck (Ed.), *Information processing in medical imaging*. Boston: Martinus Nijhoff.

Pollack, I., & Norman, D. A. (1964). Non-parameteric analysis of recognition experiments. *Psychonomic Science, 1*, 125–126.

Smith, W. D. (1995). Clarification of sensitivity measure A'. *Journal of Mathematical Psychology, 39*, 82–89.

Stewart, A., & Ord, J. K. (1991). *Kendall's advanced theory of statistics. Vol. 2. classical inference and relationship* (fifth ed.). London: Edward Arnold.

Swets, J. A. (1964). *Signal detection and recognition by human observers: contemporary readings*. New York: Wiley.

Swets, J. A. (1986a). Form of empirical ROCs in discrimination and diagnostic tasks: implications for theory and measurement of performance. *Psychological Bulletin, 99*, 181–198. (Reprinted in Swets, 1996, pp. 31–58.)

Swets, J. A. (1986b). Indices of discrimination or diagnostic accuracy: their ROCs and implied models. *Psychological Bulletin, 99*, 100–117. (Reprinted in Swets, 1996, pp. 59–96.)

Swets, J. A. (1996). *Signal detection theory and ROC analysis in psychology and diagnosis*. Mahwah, NJ: Erlbaum.

Swets, J. A., & Pickett, R. M. (1986). *Evaluation of diagnostic systems: Methods from signal detection theory*. New York: Academic Press.

Tanner, W. P., Jr. (1956). Theory of recognition. *Journal of the Acoustical Society of America, 28*, 882–888. (Reprinted in Swets, 1964.)

Wickens, T. D. (1989). *Multiway contingency tables analysis for the social sciences.* Hillsdale, NJ: Erlbaum.

Wickens, T. D. (1991). Maximum-likelihood estimation of a multivariate Gaussian rating model with excluded data. *Journal of Mathematical Psychology, 36*, 213–234.

Wickens, T. D. (1993). The analysis of contingency tables with between-subject variability. *Psychological Bulletin, 113*, 191–204.

Wickens, T. D., & Olzak, L. A. (1992). Three views of association in concurrent detection ratings. In F. G. Ashby (Ed.), *Multidimensional models for perception and cognition* (pp. 229–252). Hillsdale, NJ: Erlbaum.

Zelen, M., & Severo, N. C. (1964). Probability functions. In M. Abramowitz & I. A. Stegun (Eds.), *Handbook of mathematical functions with formulas, graphs, and mathematical tables* (pp. 925–995). Washington, DC: National Bureau of Standards.

Index

Page numbers in italic type refer to computational examples. Page numbers followed by "n" refer to footnotes. Authors listed in SMALL CAPITALS are also listed in the refrence section.